The *REST* of EUCLID

An Ancient Architecture of Arithmetic and the Modern Theory of Number

by
Robert L. Powell, Sr., Vandorn Hinnant III
and Robert L. Powell, Jr.

Published by
AV Orenda, LLC
TheRestofEuclid@gmail.com

The *REST* of Euclid: An Ancient Architecture of Arithmetic and the Modern Theory of Number by Robert L. Powell, Sr., Vandorn Hinnant III and Robert L. Powell, Jr.

ISBN 978-0-692-15930-9

Cover design: Vandorn Hinnant III
Cover image: *Window to the Golden Temple*, Vandorn Hinnant III

Illustration credits: Robert L. Powell, Sr. and Vandorn Hinnant III
The drawings by Robert L. Powell, Sr. for the original 2004 publication have been redrawn by Vandorn Hinnant III for this current release of *The* Rest *of Euclid*. All these mathematical drawings in *The* Rest *of Euclid* have been canonically constructed by hand with compass and straight edge, using the classic techniques employed by the ancient geometers.

1. Mathematics 2. Ancient Geometry 3. Fractals 4. Hidden Secrets of Number

Dedicated to Robert L. Powell, Sr.
Years 1923-2015

"Bob Powell is a clandestine national treasure
in the world of mathematics and physics."

Table of Contents

Part I.   The *REST* of 'EUCLID'

Conclusion

Part III. Some Implications of Part II of The *REST* of 'EUCLID' for 21st Century Science

Part IV. Some Implications of Part II of The *REST* of 'EUCLID' as Paradigm for a Third Millennium Politics of Ecology

Appendices

Preface

by Vandorn Hinnant III

Finally this ground-breaking thesis is being brought forth to the world. It was developed under the auspices of The Practical Science Institute, Inc. beginning in the late 1970s. By the early 2000s, the thesis had morphed through several iterations and was then finalized in 2004—and this is what is presented here.

My participation in this endeavor with Robert L. Powell, Sr. arose from an on-going rich conversation between a mathematician (Bob, Sr.) and an artist (myself) at the G. R. Lomanitz Visual Math Laboratory of The Practical Science Institute. (Houston, TX and Greensboro, NC). **This fruitful decades-long collaboration with Robert L. Powell, Sr. afforded us the opportunity of discovering an extraordinary means of performing fractal number arithmetic—with the circle.**

Here in this book are elucidations on the theorems of Euclid. The rules of Euclidean geometry are lawfully employed throughout in the constructions—which are carefully crafted following ancient canons. In deeply exploring the circle with a whole team of investigative researchers associated with the Institute, we have realized properties of the circle in Euclid's *The Elements* which have been overlooked in mathematics for centuries.

We present our findings in which we developed a way to reveal a mathematical coordinate grid system with an "in-the-plane" line segment computing system. What does that mean? There are specific points which are the geometrically lawful result of intersections of circles with circles, circles with lines, and lines with lines. Wherever a circle intersects a circle or a circle intersects a line or a line intersects a line—that intersection yields points with a specific mathematical address in the plane.

This computing system, with these canonically identifiable points permits the naming of infinitely countable square roots of one, two, three, five, etc. It also permits the precise addresses of their divisions—in other words, division by two or half of the square root of one, two, three, etc. as well as the division by three or one, two, and three thirds of the square roots of one/two/three, etc. Thus this grid enables the identification of line segments between these planar addresses (points) to infinity.

At a certain level of complexity in the plane it becomes apparent that this computing system reveals a fractal generating visual mathematical system. Other information—perpendicular, equilateral, and parallel relationships, etc.—is revealed as well as visible signs of the subtle dimensions of matter, energy, space, and time.

I encourage you to take a deep breath and enjoy the journey. Seeing does not always result in understanding. Over the years, there were many attempts made at sharing this "out of the box" thesis with others in the mathematics, physics, biology, and other scientific communities…and beyond—through direct correspondences, seminars, presentations, videos, sharing of the manuscript, et cetera.

The elegant simplicity, complexity and rigor of the thesis had succeeded in opening the "gated door"—to the mathematics of Nature and her square root numbers and their verifiable addresses on the number plane—yet it confounded many standard perspectives.

Initiating first-hand experiences became the necessary path—in order to introduce a person to the thesis. Actual hands-on drawing of the canon of the diagrams became the portal through which one could experience the "aha" of seeing what the thesis was revealing.

As in ancient times, experience is the teacher—one learns by doing.

You are invited to engage on this path of discovery through drawing the diagrams presented in this book.

Preface

by Robert L. Powell, Jr.

Dad's explorations into sacred geometry began in the mid 1970s with an inquiry from John Biggers, the founding chair of the art department at Texas Southern University and a noted African-American muralist. Mr. Biggers was looking for someone to investigate "an African mathematics" with him. They identified a set of geometric patterns that are created with the use of a compass and a straight edge used by the ancient Egyptians as well as other builders, artists and scientists throughout history. Biggers incorporated these learnings as "his work became more geometric, stylized and symbolic"[1]. For Dad, this collaboration initiated a 30-year exploration of drawing, reflections, analysis and sharing. He drew constantly. He devoted the craft, creativity and focused attention to these exercises he had developed during his career as a research physicist.

The drawings in this book are the tip of the iceberg of the thousands of times he sat at his board at his home or mine, visiting friends, collaborating with Vandorn or conducting "Boot camp" sessions. More than viewing these representations, the experience of observing Dad or better yet participating in the creation of everything from the simplest constructions of "first 12 pages" during a boot camp to the lesson Nelson Stover created for school kids to the drawings you see in this book offered the opportunity for us to enjoy the metaphysical aspects of this work as Dad would describe it as well as receiving the benefits to head, hand and heart of doing the work.

An example of Dad's metaphysical take on the geometry is the fact that you only physically see the centers of the initial three circles once the third circle is drawn. Dad would say that "3 is the first number". This book represents collaboration of three individuals, Dad, Vandorn and myself. The interactions of our three circles span art, architecture, science and law. A key to this book is the understanding of graphical representations of mathematical concepts the three of us share. At the same time, the numerical formulas and concepts are all Robert Powell, Sr. Dad did not refer to himself as a mathematician since his freshman year at Fisk. However, his entire professional career revolved around mathematical representations of 3-D physical space and the forces at play in it.

The geometry he explored was an extension of his lifelong work. I trust the readers of this book may find in this book insights to help with your investigations of the graphical, numerical and metaphysical nature of things.

---

1   Eglash, Ron (2004). "A Geometrical Bridge Across the Middle Passage: Mathematics in the Art of John Biggers." The International Review of African American Art, Vol. 19, no. 3, pp. 29-33.

About this Book...

In his extraordinary and rigorous approach in this study, Robert L. Powell, Sr. focused on accessing the ancient canons of geometric principles as well as immersing himself in mathematic and scientific theories and applications found throughout the centuries and up to our modern era. With these elements gestating within him, Robert L. Powell, Sr.—who was already steeped in mathematics, physics, holography and other scientific disciplines—investigated for decades how to canonically bring forth a contemporary method of inquiry as well as a tool of precise mathematical "languaging" for the numbers of Nature.

He would often lament that at the turn of the century—meaning at the beginning of the 20th century—an international consortium of professional mathematicians created what he referred to as "a gated community" by formally excluding the body of square root numbers as those numbers that were "irrational" and "did not have proper addresses" on the number line.

Perhaps it took Robert L. Powell, Sr. being highly perspicacious and a person of color—and his team—to see through the folly of that and to eventually and ingeniously determine that with *The* REST *of 'Euclid'* one could reclaim those irrational numbers and find their addresses by way of THE NUMBER PLANE!

Throughout these investigations, Robert L. Powell, Sr. unceasingly attempted to encourage his scientific colleagues—in physics, biochemistry, biology, mathematics, etc.—to understand that **these irrational, square root numbers ARE the NUMBERS OF NATURE—that they ARE KEY for understanding and interpreting our world throughout the physical and subtle spectrums—from the micro to the macro.** Instead of computing rows and rows of calculations lined up after the proverbial "decimal point"—he insisted that the elegance and precision of the square root numbers reveal a deeper level of structure and of resolution. He completed this manuscript in 2004 and shared it widely with colleagues.

ADVISORY: Robert L. Powell, Sr. drafted this 2004 manuscript in his own inimitable style with its intentional capitalizations, italics, brackets, footnotes, formatting, etc. Even though editorial help was offered he insisted to leave it all be. His instructions were to publish the manuscript "AS IS." So thus it is.

Foreword

By Robert L. Powell, Sr.

1. WE ARE a physicist, an architect, and a _visual_ artist—a trans-disciplinary team of non-mathematician _intellectuals_. Our connection is a complementary set of professional interests in the study of a _trans-intellectual_ question: by what methods were the ancient guilds of architects and artists so able to embody into their works the _gestalt_ element of beauty and fitness that characterizes their agelessly elegant compositions?

   A quite adequate capture of the 'flavor' of the elusive aesthetic property we study can be inferred from our paraphrase of a set of remarks by Professor Dan Pedoe—himself a careful student of the guilds. Early in his _pleasure_-encouraging book, _GEOMETRY and the VISUAL ARTS_, Pedoe gives a review of _the appropriate breadth and depth of concern with the whole subject of architecture_, as recommended by Marcus Vitruvius—himself one of the significant links in our unbroken chain of inherited guildsmen[2]

   Our paraphrase:

   > The elegant construction depends on _order, arrangement, eurhythmy, propriety, symmetry, and economy._ ...Order gives due measure to the members of a work considered separately. Symmetry gives agreement to the proportions of the whole. It is an adjustment in according to quantity. By this is meant the selection of modules from the members of the work itself, and constructing the whole work to correspond. Eurhythmy is beauty and fitness in the adjustment of the members of a work.
   >
   > [_...and..._]
   >
   > _Symmetry_ **is** a proper agreement between the members of the work itself, and the relation between the different parts of the whole general scheme, _in accordance with a certain part **selected**_ as **the standard**.

   Any search for the technical methods by which these ineffable aesthetic embodiments into _visual_ structures are accomplished _must_ lead an interest such as ours to a study of the guilds' _strategic_ employment of their three essential instruments for producing geometric constructions in _the plane_:

   the compass;
   the straight edge;
   and [!!!]
   an _un-marked_ plane surface, of some sort.

Our experimental studies of the ancients' works, using the *three* essential tools, compelled us to consider, as a working proposition, that the guildsmen were governed rigorously by a protocol of rules which nevertheless permitted, perhaps indeed *guaranteed,* the consequent embodiment of the ineffable *qualities* captured in the Vitruvius précis. The remark "...in accordance with a certain part *selected* as *the* standard" was particularly compelling as an organizing clue to guide our experimental strategies at de-constructing[3], and re-constructing, the Implicate Order plane geometry patterns which seemed to govern the general scheme of the evolution of a work.

Indeed, this snippet from Vitruvius was found to explain *precisely* the key to the success of our program of gestalt pattern analysis and synthesis. In over a couple of decades of playful study, we slowly learned that the ancients merely *required* that their works be *Theorems* of *Plane* and *3-dimensional Geometry*!!

That is to say, the Implicate Order *pattern* by which the guilds assured the gestalt coherence of the whole general scheme was an Implicate Order *palimpsest* determined strictly by the Five 'Rules of Euclid' which govern and assure the construction of a Theorem.

We recognized that this system of rules is the *complete* set of Canons that *governed* and *guided* their choices of composition strategies and tactics for relating between the different parts of the whole general scheme, *in accordance with a certain part selected as a reference.*

We recognized that centuries of canonical constructions—from ancient Egypt's Temple of Karnak to New York City's Cathedral of Saint John the Divine—give enduring record of the guilds' adherence to the perennial accumulation of a syllabus of strategic, canonical, gestalt-embodying palimpsests that are governed by this protocol.

2. This SET OF (FIVE) RULES comprises three of the five 'Postulates of Euclid' which together with the five 'Axioms of Euclid' provide the platform for Euclid's masterpiece, *The Elements.* One of those unique books like the Bible which seem to fuse the best efforts of generations of creative minds into a single inspired, creative whole, *The Elements* is a work of such commanding lucidity and style that some scholars consider it the most coherent collection of closely reasoned thoughts ever set down by humans.

    The Elements contains 13 books, or chapters, which describe and prove a good part of all that the human race knows, even now, about lines, points, circles and the elementary three-dimensional shapes.[4]

3. Our program of analysis and synthesis of the ancient palimpsest structures has permitted the recognition of a related pair of Theorems which is the foundation for the construction of *all possible* Implicate Order patterns embodying the gestalt eurhythmy, symmetry, and economy. This inter-articulated pair of theorems extends, qualitatively and quantitatively, what the human race

knows about lines, points, circles and the elementary three-dimensional shapes. Our extension of this knowledge, our ancient pair of theorems, emerges as implications embodied in the teachings of *the very first Problem* of *the very first Book* of *The Elements*.

To our great surprise, the pair of theorems <u>*also*</u> provides—permits the definition of—a remarkable *generalization* of the <u>foundation</u> of the <u>architecture</u> of the <u>structure</u> of our modern mathematics. The revolutionary generalization also, of course, *evolutionizes* applied mathematics as a 3rd millennium scientific instrument.

We refer to this ancient pair of implications of Proposition 1, Book I—ignored, neglected, over looked for two millennia now—as *The* REST *of 'Euclid'*.

4. The purpose of our book, then, is two-fold: (1) to present this serendipitous discovery of a *revolutionary* generalization of the elementary mathematical concept, Number, and its co-related *evolutionary* mathematical architecture, an *ab inito* <u>non</u>-linear Arithmetic; and (2) a cursory sketch of the *evolutionary* applied mathematics instrumentations co-related to the richer Arithmetic.

5. For cultural reasons, the professional mathematician must, as a conditioned reflex, play Devil's Advocate to the claim of discovery by a trio of non-mathematician amateurs of such a humongous un-tapped epistemological universe for the Western Europe intellect as:

    (1) our *The* REST *of 'Euclid'* and
    (2) its co-related evolutionary applied mathematics enablements.

    Surely the 'West'-dominated guild of mathematicians has already *fully* exhausted the Teachings of *The Elements* as a rational source of what is possible to know.

    Therefore, so as not to be dismissed peremptorily by the likes of Ian Stewart and John Casti as a 'Math Crank'[5], our book's presentation has tried to adhere to that profession's arcane (and in key places, now obsolescent) vocabulary, grammar and rhetoric. Nevertheless, we intend the interested non-mathematician reader to be able to grok the simple revolutionary and evolutionary features of the generalizations. Hence the presentation seeks also to subvert the conditioned reflex aversion-to-mathematics of the non-mathematician reader.

---

2   Pedoe, Dan. *Geometry and the Visual Arts*. New York; Dover 1983. p. 18.
3   Derrida, J. *Edmund Husserl's Origin of Geometry: An Introduction*. Stoney Brook: Nicholas Hays, Ltd. 1978.
4   David Bergamini, et al [Rene Dubos, Henry Margenau, C. P. Snow], eds. *Life Science Library Mathematics*. New York: *Time Incorporated*. 1963. p.45, ff.
5   George Johnson. "Genius or Gibberish? The Strange World of the Math Crank," *New York Times*, Tues., Feb. 9, 1999, p. D-1.

Introduction

In PART I

A. **The <u>Act</u> of <u>constructing</u>** a theorem in plane geometry *requires* the cooperative interplay between three entities:

(1) the **Hardware** – the 'circumference'-maker; the 'radius'-maker; and an un-marked plane surface on which to inscribe rules-allowed circles and rules-allowed radii;

(2) the **Software** – the 'SET OF FIVE RULES' which govern the [theorem-maker's] freedom to lay down permissible circumferences, radii, and radius-extensions on the plane; <u>***and***</u>

(3) the **Humanware** – the intellectual craftiness [of the theorem-maker] to extract the maximum of Theorem 'Intelligence' permitted by the **Hardware/Software** combination.

Note: The term Humanware refers to the human being who is actively engaged in the thinking process as the cognitive "theorem-maker" in the canonical operation.

The *action*, itself, is a *process* by which the theorem-maker <u>sequentially</u> accumulates a coherently related *system* of rules-permitted circumferences, radii, and radius-extensions. At any point in the process the **Humanware** is provided a *visual*, <u>parallel-organized</u>, two-dimensional 'print-out' of both (a) the coherent system of circumferences and lines and (b) the point-set of circle-circle, circle-line, and line-line intersections: the synthesized *complete set* of solutions to the 'simultaneous'-organized system of canonical non-linear equations.

Since 'forever ago' pre-Europeans and our own guilds of Sacred Architecture and Sacred Art have guaranteed the mathematical coherence of the parallel organized system of such palimpsest equations by the **Humanware's** requirement that the entire parallel-organized 'Theorem' be developed *in accordance with a certain part <u>selected</u> as <u>**the**</u> standard*.

B. For the ***selection as <u>the</u> standard***, relative to which the coherence of the parallel-organized 'Theorem' is constructed, the **Humanware** introduces into the *otherwise un-marked plane* (Figure 1) an arbitrarily positioned *pair* of <u>*points*</u> (Figure 2).

POWELL, HINNANT, AND POWELL

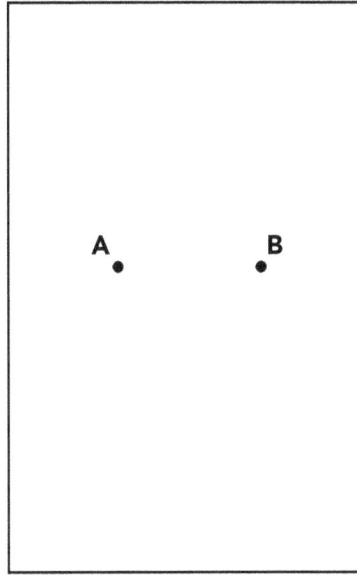

Figure 1. An _Un_-marked Euclidean _plane_

Figure 2. A reference binary pair, [A,B], giving coordinate calibration to the _entire_ Euclidean _number plane_

This plane initializing *binary point system* permits sequential introduction into the plane, by the Humanware, of the five implications of three of the five 'Postulates of Euclid'.

The five independent implications are shown in Figure 3a & Figure 3b and in Figures 3c, d, & e.

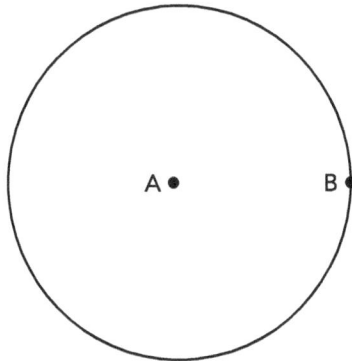

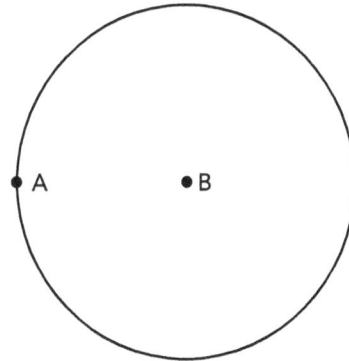

Figure 3a.

Figure 3b.

The five canonical operations permitted by any Binary Operand.

Figures 3a & 3b. _The_ pair of circles that _share_ the radius, determined by the reference point-pair [A,B]; _and_ which *specify* which plane is referenced. Figures 3c, 3d, & 3e. The line segment, and its pair of extensions, determined by the reference point-pair, [A,B].

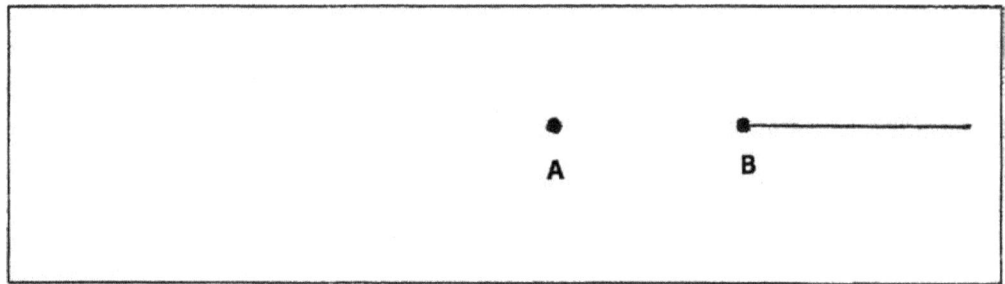

Figure 3c.

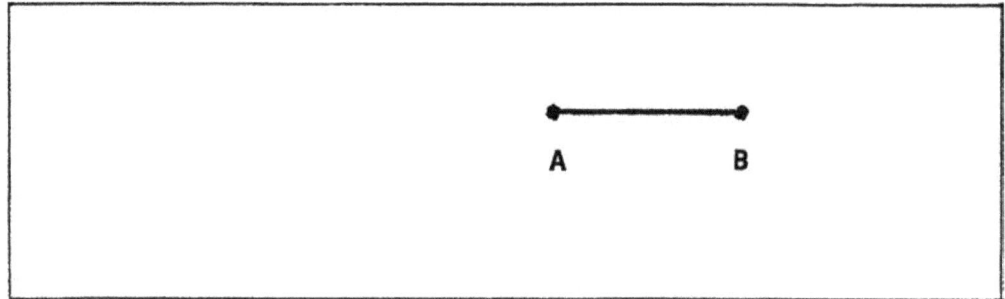

Figure 3d.

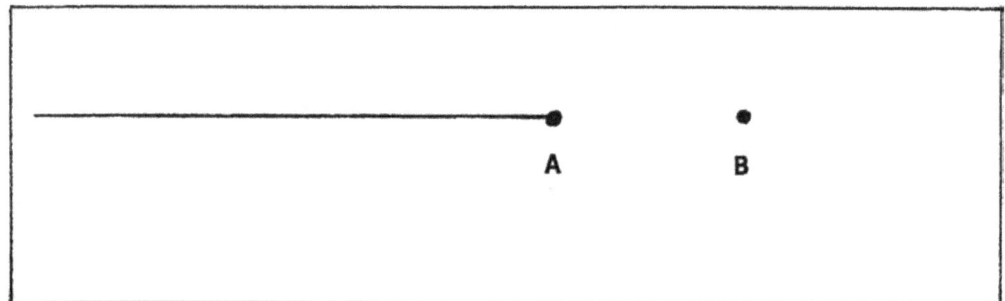

Figure 3e.

The *complete, planar,* parallel-organized implication of the <u>reference</u> <u>standard</u> binary point system is shown accumulated in Figure 4.

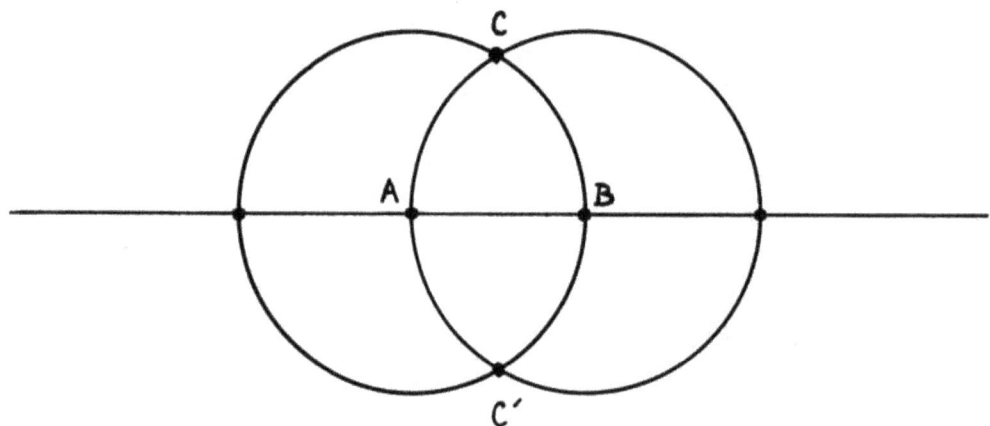

Figure 4.

First and second millennium mathematics is constructed on the foundation of an 'Elementary' Number Theory based on the one-dimensional Number <u>Line</u> calibrations shown on the extensions of the segment AB, in Figure 4.

In this book we make ***complete*** use of the epistemologically significant Teachings of THE *REST* OF the implications of Figure 4.

In particular, the *pair of circumferences* with *a shared radius*—permitted by the *reference point-pair binary*—will be exploited in Chapter I to develop a Theorem of <u>plane</u> geometry which defines a hierarchy of non-'Elementary' Number Theories, within which our modern 'Elementary' Number Theory is *imbedded*, as an almost empty proper sub-set.

This theorem development will require the presence of the 'Green' circle pair shown in Figure 6. A table of fractal number Trigonometry values is generated as a self-replicating corollary.

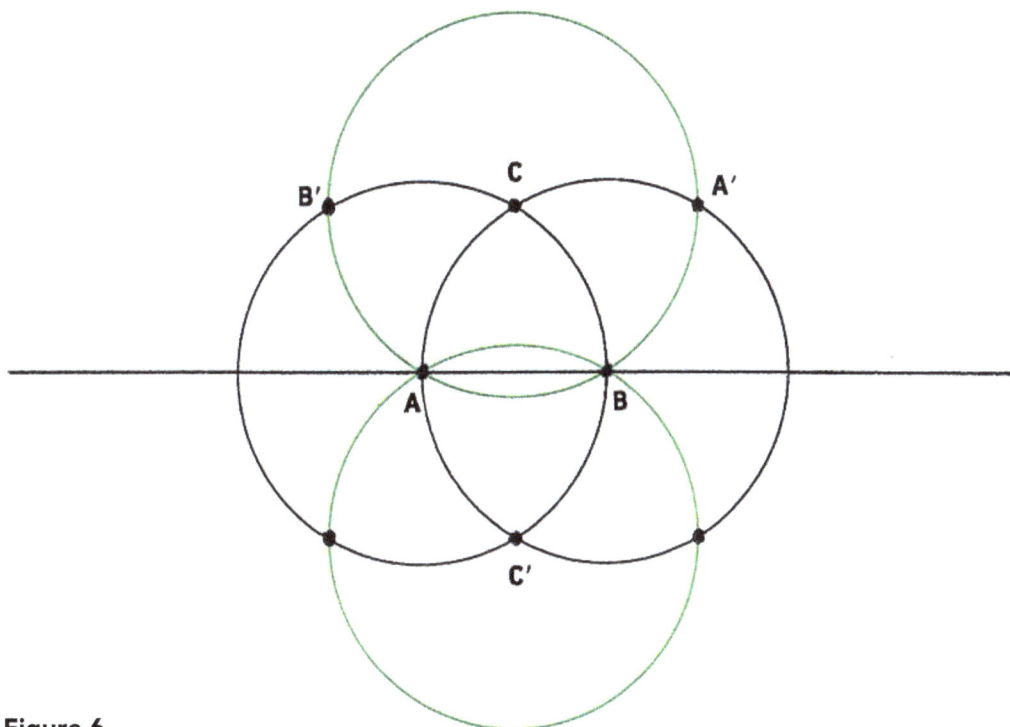

Figure 6.

In Chapter II we 'break the symmetry' of the Green circle-pair, to develop a hierarchy of three Theorems which calibrate the entire [Euclidean] plane in a regular lattice of positional notation coordinate systems, in terms of the powers of the three fractal integer base numbers. Tables of fractal number-based logarithm numbers are generated as self-replicating to-scale corollaries.

In Chapter III we break the symmetry of Figure 3a (or 3b), to develop a Theorem which gives plausibility to the remaining two of Euclid's Five Postulates.[6]

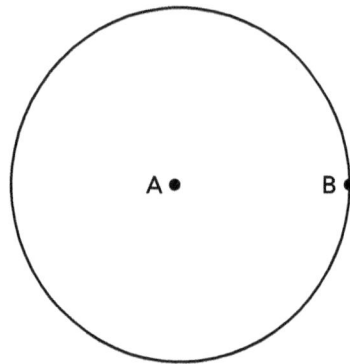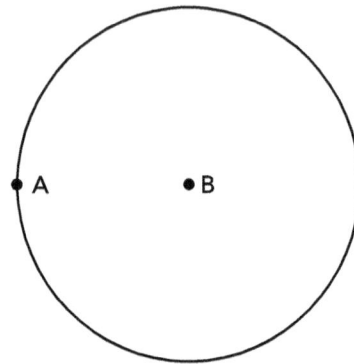

Figure 3a.   **Figure 3b.**

Chapter IV is a summary of the *evolutionary* implications for ['pure'] mathematics of

The generalization of our Procrustean concept of the mathematical notion, Number: The fractal integer number *interpolations* on the Number 'line':

The three space-filling fractal number ratios for calibrating the Number <u>plane</u>;
Fractal number kinematics for the plane;
The group theoretic property of Number;
The planar geometry of the Transcendental Number integers and their
    cyclotomic eigenvalues;

Addressing the limitations

… of our Procrustean concept of Arithmetic, Analytic Geometry, Kinematics, & Trigonometry.

… of our Procrustean concept of the notion, Algebra:
        synthesis and analysis;
        geometric (Clifford) algebra;

… of our Procrustean concept of the notion, Analysis:
        analyticity, and the quantization of violation of analyticity;

In PART II

Applications of this Approach will be addressed, including:

Implications of the fractal number architecture for 21st century mathematics

Implications of the circumference number integers and their fractal number integers for our 21st century theory of applied mathematics.

---

6   David Bergamini, et al [Rene Dubos, Henry Margenau, C. P. Snow], eds. Life Science Library *Mathematics*. New York: *Time Incorporated*. 1963. p.46.

PART I.
The *REST* of 'EUCLID'

Chapter I. DEFINITION OF THE <u>Non</u>-ELEMENTARY NUMBER THEORY

A. Figure 4 expresses the full implication of the reference location and orientation of the binary point-pair, AB, selected as the standard calibration of an otherwise un-marked plane, Figure 1. This complex of canonical implications is seen to consist of:

(1) a symmetry-related circle-pair, having a shared radius [the radius is specified by the reference point-pair],
(2) a symmetry-related pair of extensions of the shared radius, and
(3) the *complex* of *point-pairs* consisting of an articulated pair of symmetry-related point-pairs.

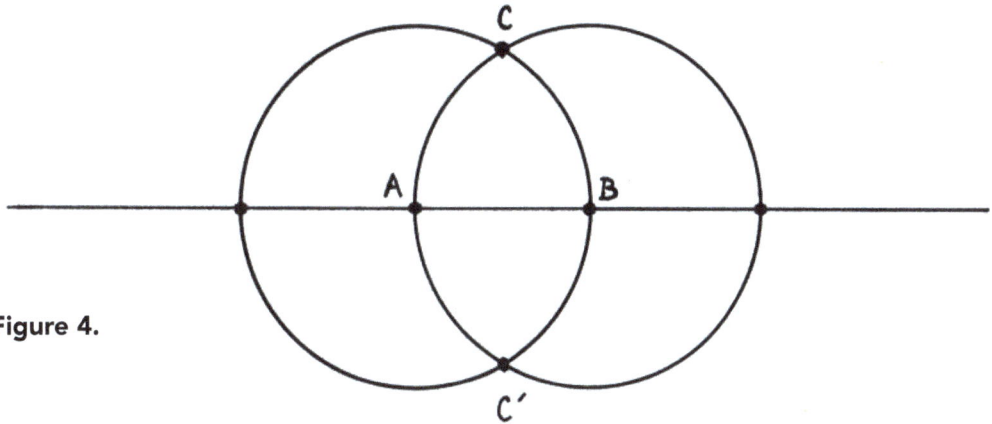

Figure 4.

For the Theorem-developing strategy the Humanware seeks to carry out in this chapter, we augment this Figure 4 with the mutually symmetric green circle-pair shown in Figure 6.

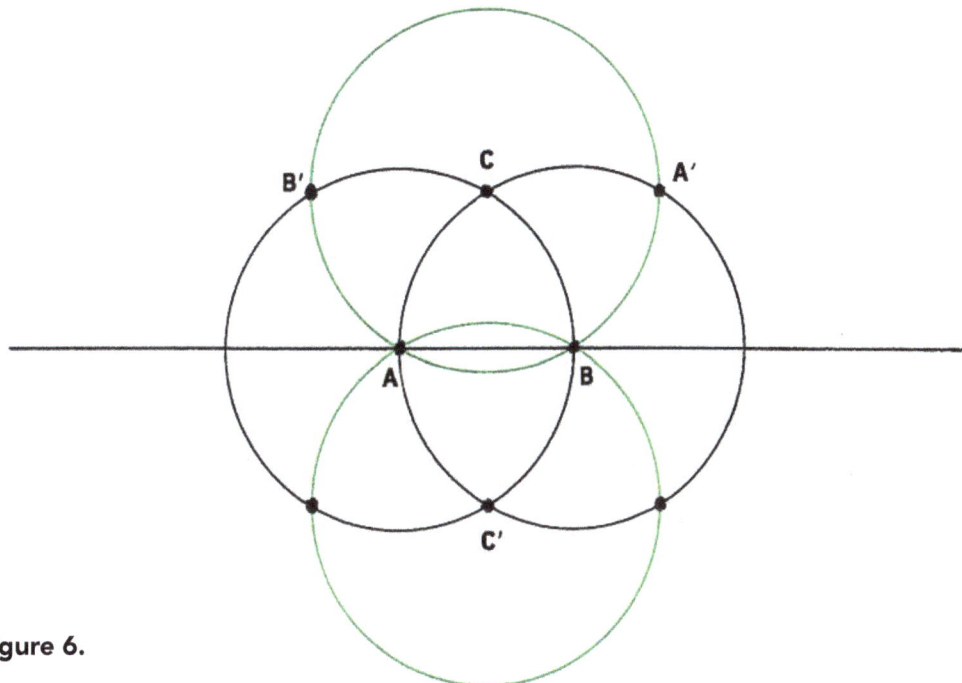

Figure 6.

The green circle pair's articulation with the black circle-pair, in Figure 6, permits the construction of the pair of black line parallel pairs in Figure 7. This pair of parallel lines creates the quartet of new points, E, F, G, and H.

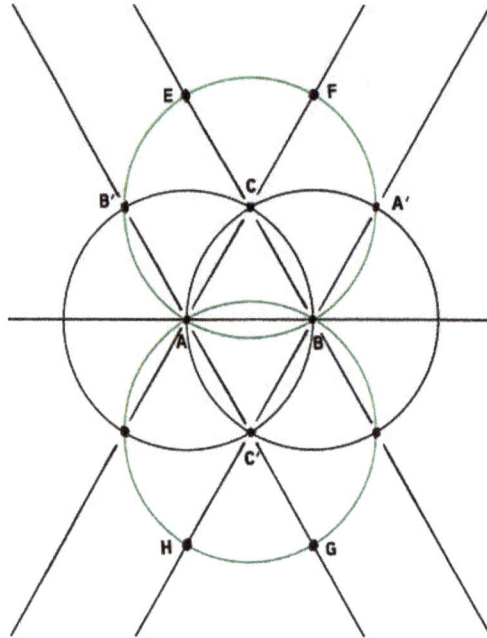

Figure 7.

The quartet of new points permits construction of the green line-pair shown in Figure 8. The green line-pair creates the quartet *pattern* of *points*, I, J, K, and L.

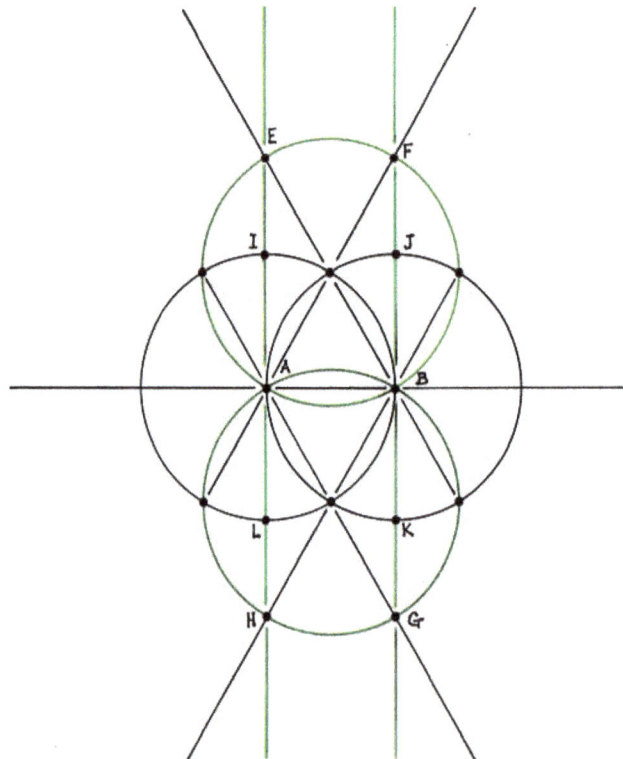

Figure 8.

B. The complex of point-pair patterns in Figure 8 permits the _selection_ of a pair of pairs of radii for a significant pair of new circles:
   (1) the _green_ radius pair AA' and BB' and
   (2) the _blue_ radius pair AJ and BI, shown in Figure 9.

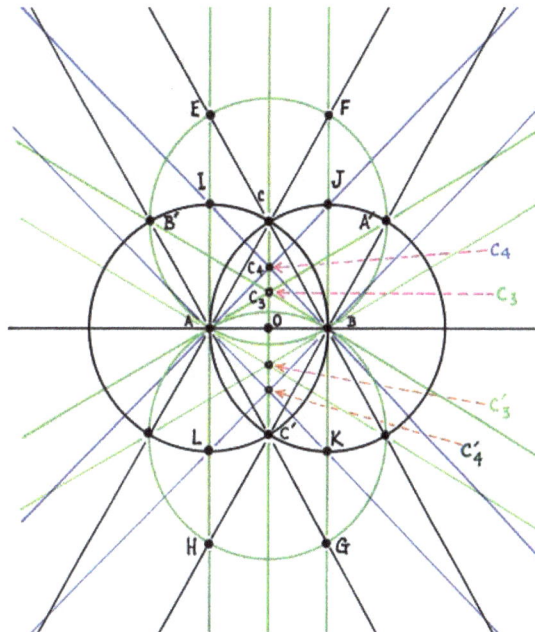

Figure 9.

This pair of pairs of radii permits construction of the pair of blue circles and the pair of green circles shown in Figure 10a.

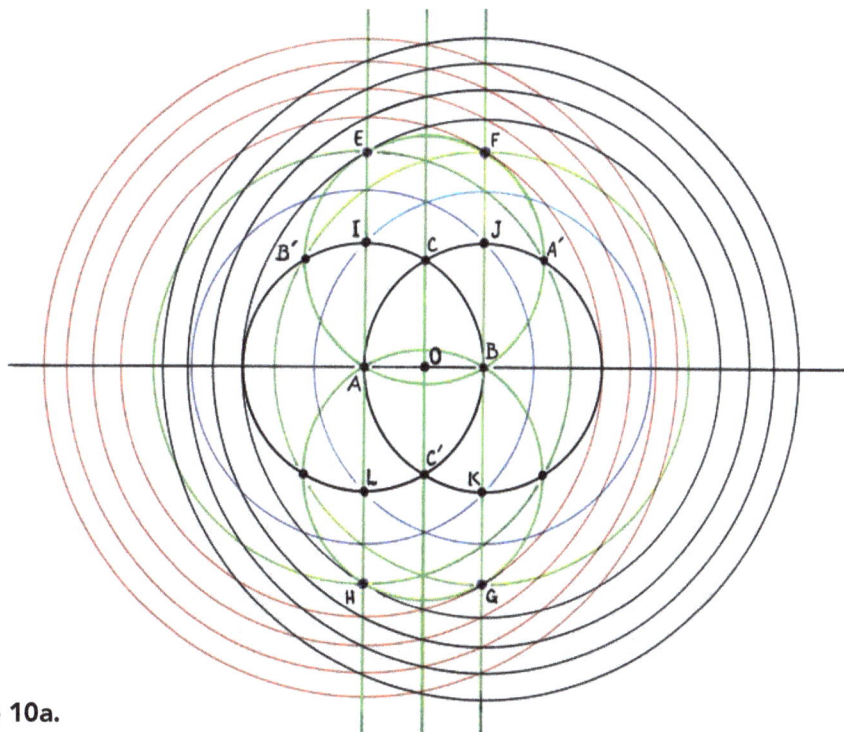

Figure 10a.

The blue circle-pair is the first scaled replication of the black circle-pair. The green circle pair, a scaled replication of the blue circle-pair, is the second scaled replication of the black circle-pair... .

3

> **HIATUS:**
> In Chapter III we examine the symmetry relation between the green line system and the black line. There we determine that each green line is *necessarily* '*perpendicular*' to the black line.
>   A corollary of this symmetry relation is a Non-linear Addition relationship (the 'Pythagorean' theorem) which connects successive radii in a countably infinite system of scaled replications of the black circle-pair.
>   For this chapter, however, we assume provisionally that each green line is 'perpendicular' to the black line.
>   We make use of the corollary Non-linear Addition relationship to <u>define</u> the radius r_{n+1}, of the $(n+1)^{th}$ circle, as the scaled increment of the radius r_n, of the nth circle, the increment being determined by the shared radius r_1, of the black circle-pair:
>
> $$[r_{(n+1)}]^2 = [r_n]^2 + [r_1]^2 \qquad \text{(an old quadratic sum rule)}$$

C. Figure 10a shows an articulated pair of concentric circle families. Assuming perpendicularity between each green line and the black line, then, in each family, the radius of the $(n+1)^{th}$ circle is <u>determined</u> as the hypotenuse of a right triangle having sides r_1 and r_n.

D. As foundation for the modern theory of number, we <u>***choose***</u> Figure 4 to be the standard *planar, <u>vector</u> reference <u>complex</u>*, with its strategic measure assignment:

$$AB = \sqrt{1} \qquad\qquad\qquad [1]^*$$

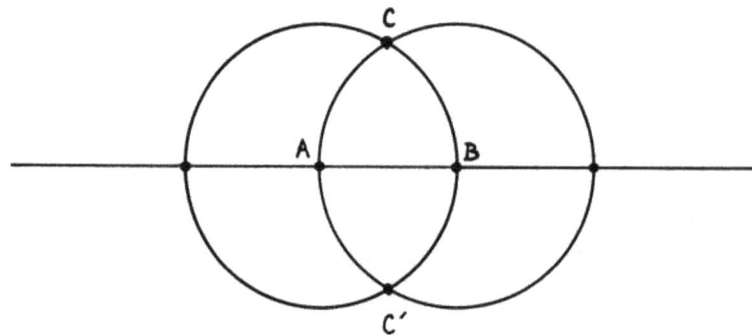

Figure 4.

With this selection for the standard reference <u>length</u>,
  AJ (in Figure 10a) is the hypotenuse of the right triangle with sides $= \sqrt{1}$ and $\sqrt{1}$. This computes, for AJ, the radii of the blue circle-pair:

$$AJ = \sqrt{2}. \qquad\qquad\qquad [2]$$

  AA′, the radius of the green circle-pair is seen to be the hypotenuse of a right triangle with sides $AB = \sqrt{1}$ and $AJ = \sqrt{2}$. This computes for the hypotenuse AA′, the radii of the green circle pair:

$$AA' = \sqrt{3}. \qquad\qquad\qquad [3]$$

The radius of this third circle-pair, on a green line, and the shared radius of the first circle-pair compute a hypotenuse which generates the radii of a fourth circle-pair, with radius r_4:

$$r_4 = \sqrt{4} \qquad\qquad\qquad [4]$$

$$\ldots$$

$$r_{[n+1]} = \sqrt{\{[\sqrt{1}]^2 + [\sqrt{n}]^2\}}$$
$$= \sqrt{[n+1]}, \quad n = 0, 1, 2, 3, \ldots \qquad [5]$$

E. The _complex_ of circles, extended diameters, and points accumulated in Figure 10a constructs the _architecture_ for the generator of a _countably infinite set_ of mathematizable properties of _the circle_.

Characteristic number values are selected, by the generator, for:

(1) the circle area,
(2) the circle circumference, and
(3) the circle diameter/radius. These number values—transcendental number _integer_ values for (1) and (2): 'trans-rational' number _integer_ values for (3)—are determined by the measure assignment, relationship [see 1].

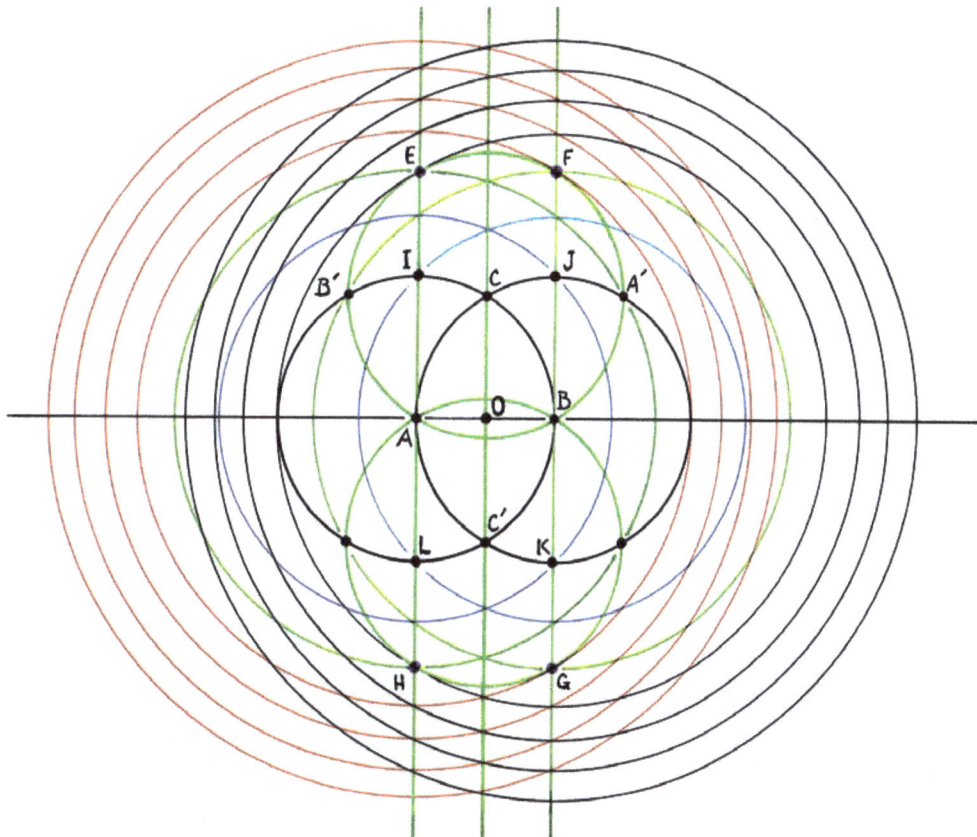

Figure 10a.

Figure 10b shows the result of the Induction Process recipe for hypotenuse generation of the (n+1)th radius, of the countably infinite set of circles.

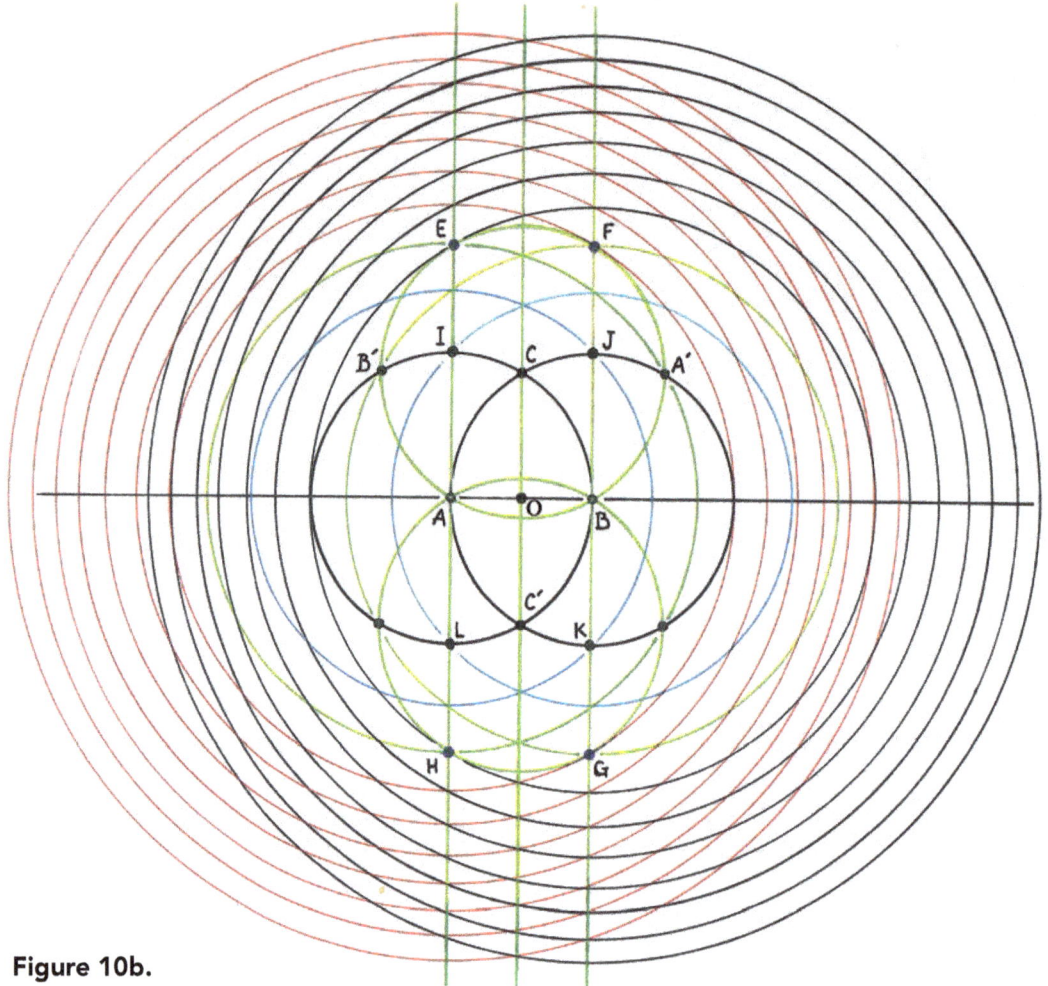

Figure 10b.

The transcendental number integers and 'trans-rational' number integers thus generated, inductively, as properties of the nth *circle*-pair are:

$$\text{Area}_n = \pi(r_n)^2 = \pi n, \qquad [6]$$

$$\text{Circumference}_n = \pi(2r_n) = 2\pi\sqrt{n}, \qquad [7]$$

$$\text{Radius}_n = \sqrt{n}. \qquad [8]$$

Values of [7] and [8] assume special values for the almost empty sub-set of n-values,

$$n = 4, 9, 16, 25, 36, 49, \ldots.$$

Each of the three elementary properties, [6], [7], and [8], supports the definition of a Euclid-based Number system, with its associated Euclid-based Arithmetic.

F. The *complex* of concentric circle family-pair, articulated by the reference line segment and the three green lines in Figure 10b constitutes, moreover, a canonical system of linear and quadratic equations *and* a *complete* set of solutions to the total system of equations. This set of canonical algebraic solutions implicit in 'Euclid' will be explicated in Chapter IV. There the patterned *self-replication, to-scale,* system of solutions will be shown to constitute a two-dimensional, countably infinite Table of fractal integer number Trigonometric ratios, a here-to-fore *un*-reported mathematical intelligence implicit in Euclidean geometry.

G. The strategy directing the sequential steps to generate the pair of families of concentric circles in Figure 10b from the pair of circles which share the reference AB as a common radius is describable as a *symmetry-breaking* process: the breaking of the shared radius property, unique to the reference binary pair, is systematically increased, to-a-scale, retaining the common centers.

Chapter II.   FRACTAL NUMBER INTEGER ANALYTIC GEOMETRY

A. The green circle pair articulation in Figure 6 embodies a certain unique spatial articulation relationship to the standard reference segment AB: (1) the radius of the green circle-pair *equals* this length; and (2) the green circle pair *shares* this reference length, as a common *chord*.

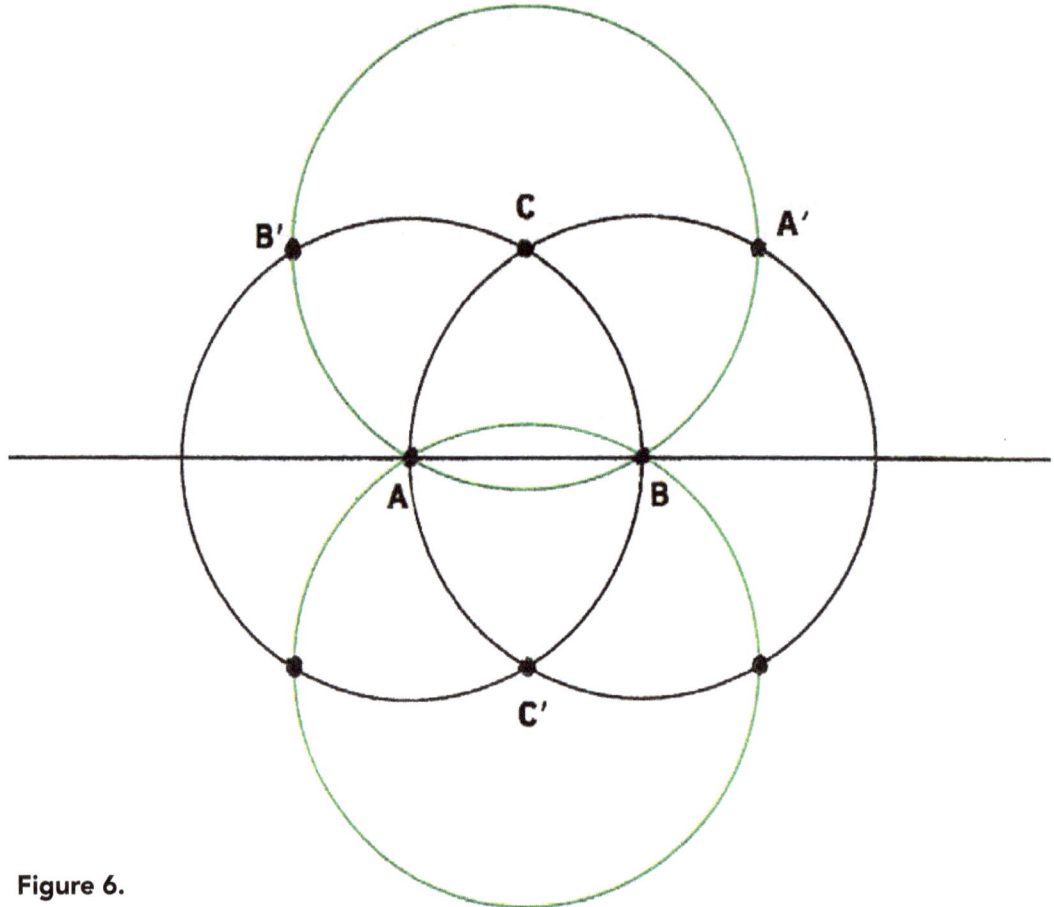

Figure 6.

Guided by analysis and synthesis of ancient geometric pattern constructions, we construct a hierarchy of three canonical algorithms which lead to periodic lattices of fractal number integer calibrations of the Euclidean *number plane*.

The most primitive number calibration of the periodic lattice of planar coordinates will be in terms of positional notation in the ± powers of the fractal number integer, $\sqrt{3}$.

This calibration permits derivation of a calibration of the Euclidean *number plane* in terms of a periodic lattice of planar coordinates in terms of positional notation in the ± powers of the composite fractal number integer, $\sqrt{2}$, and a third derivation of a calibration of the plane in terms of a periodic lattice of (orthogonal) coordinate systems in terms of positional notations in the powers of the composite fractal number integer pair, $\frac{1}{2}[\sqrt{5} \pm \sqrt{1}]$.

This hierarchy of *plane-filling* coordinate systems is characterized by a sequence of systematic <u>breakings</u> of the symmetry relation, (1), in Figure 6, while preserving the reference chord relation, (2).

We begin this symmetry-breaking sequence with the platform *complex*, Figure 9.

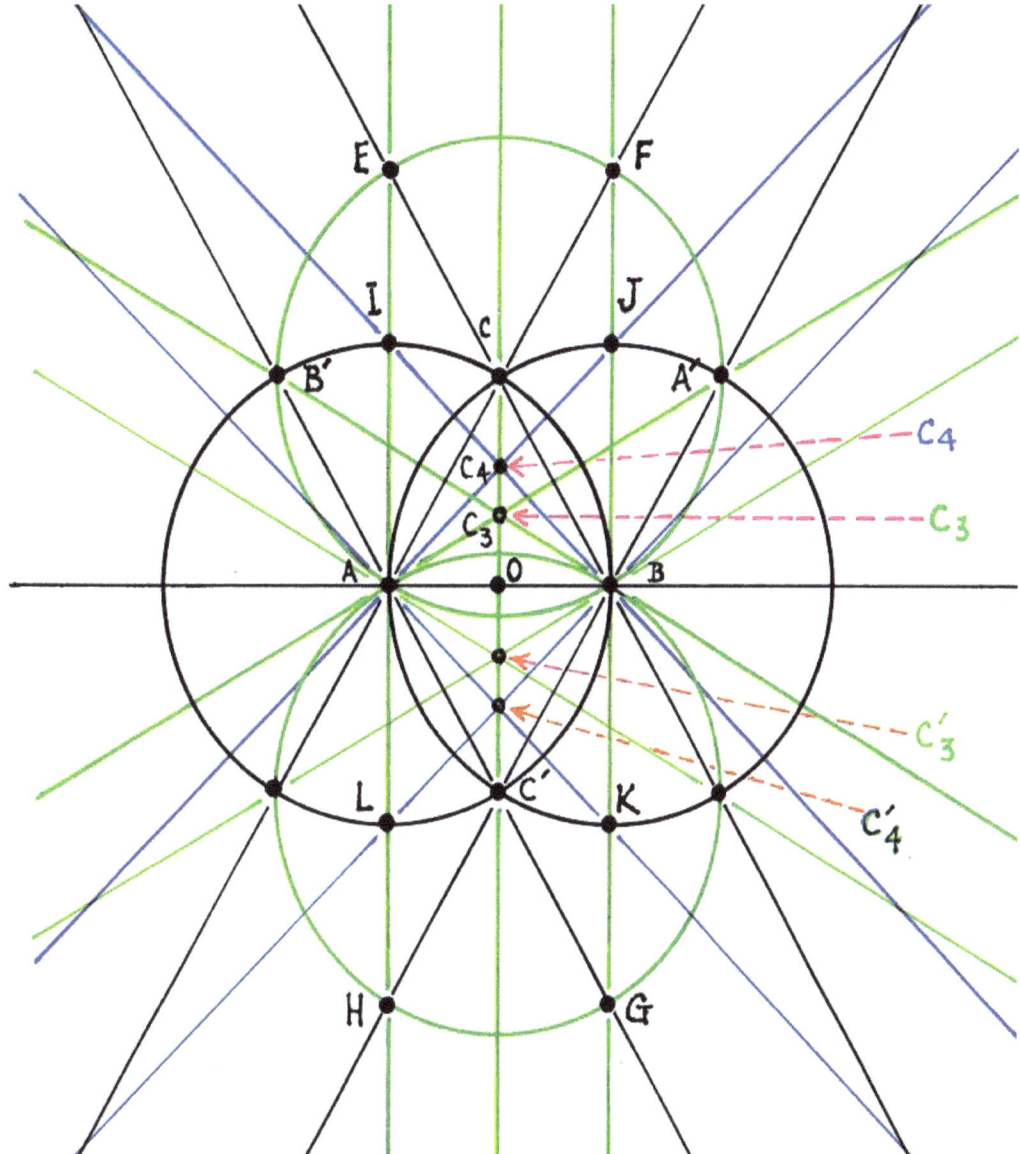

Figure 9.

B.1. In Figure 9 the three green line segments, AA′, BB′, and CC′, produce the new point of their common intersection, C_3.

With C_3 as center, construct a green circle, with radius AC_3. (See Figure 11.) Construct the image 'partner' of this green circle, below the reference segment, as shown in the Figure 11: the green circle, centered at C'_3 with radius AC'_3

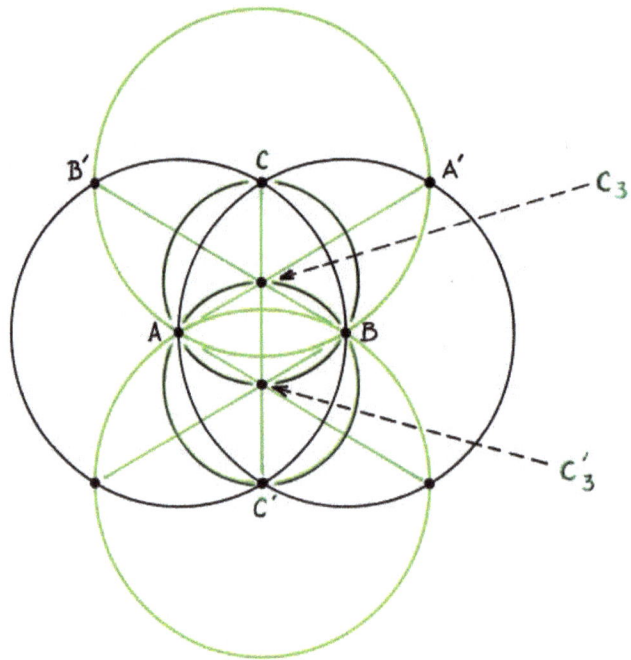

Figure 11.

2. In Figure 9 the two blue line segments, AJ and BI intersect, producing the point C_4. In Figure 12, with C_4 as center, construct a blue circle with radius AC_4. Construct the image partner of this blue circle, below the reference segment, as shown in the Figure 12: the blue circle, centered at C'_4 with radius AC'_4.

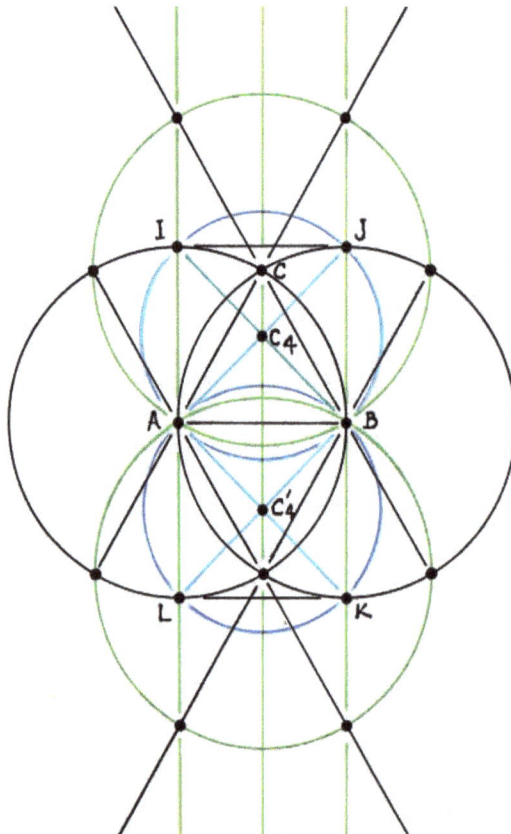

Figure 12.

3. The quartet of points, I, J, K, & L in Figure 12 also permits construction of the pair of orange segments, IK and JL, in Figure 13. The intersection of the orange segments produces the point, O, in Figure 13.

The orange circle in Figure 13 is produced with center, point O, and radius OI. The orange circle constructs the point-pair, M & N, on the reference segment extension.

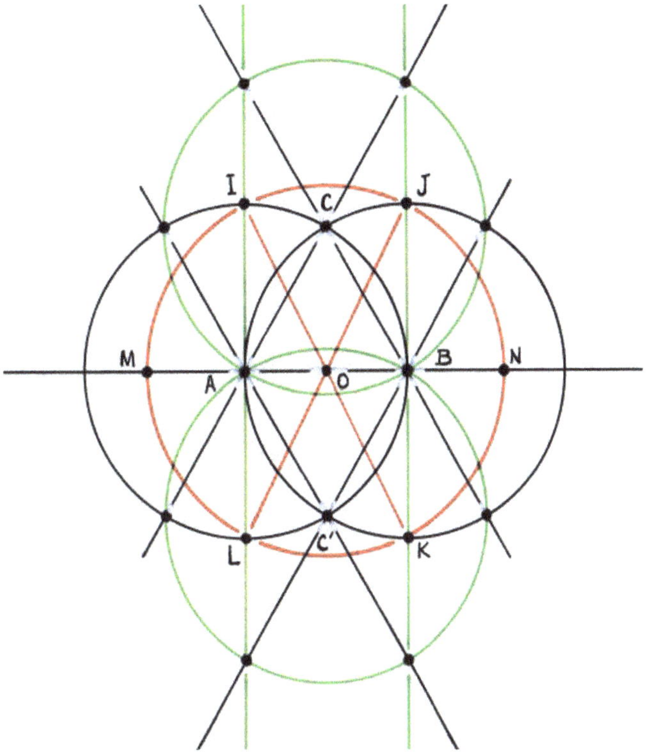

Figure 13. The 'Orange' circle, radii ½[IK]

The point-pair, M & N, permits construction of the purple circle-pair in Figure 14, with a center at the A point, radius AN and with a center at the B point, radius BM.

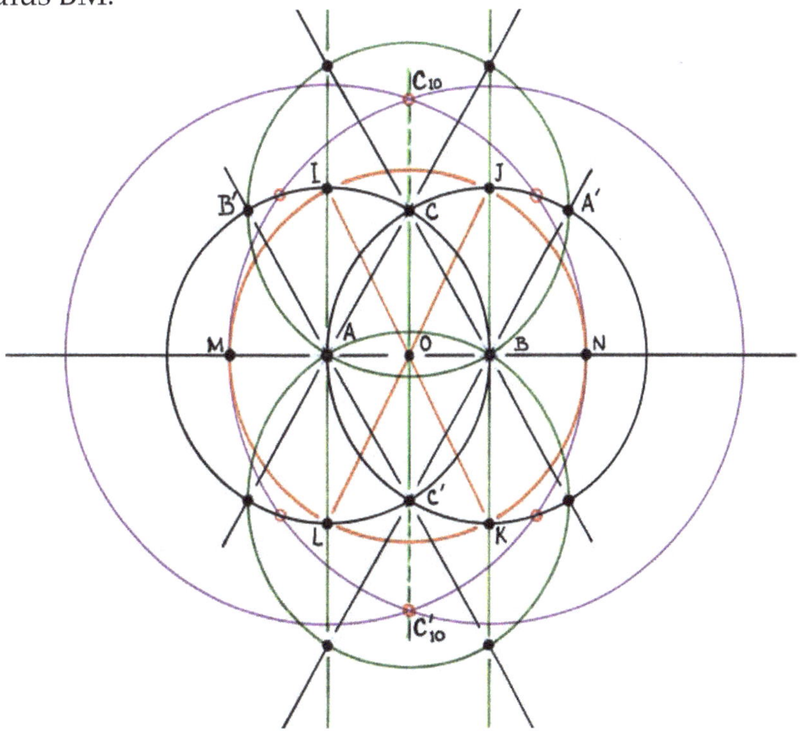

Figure 14. Purple circle-pair, radii = Φ[=½(...+...)]

Using the purple circle-pair construction from Figure 14…
  the intersections of the purple circle-pair, C_{10} & C'_{10}
  are centers of the black circle-pair in Figure 15, with radii $C_{10}A$ & $C'_{10}A$.

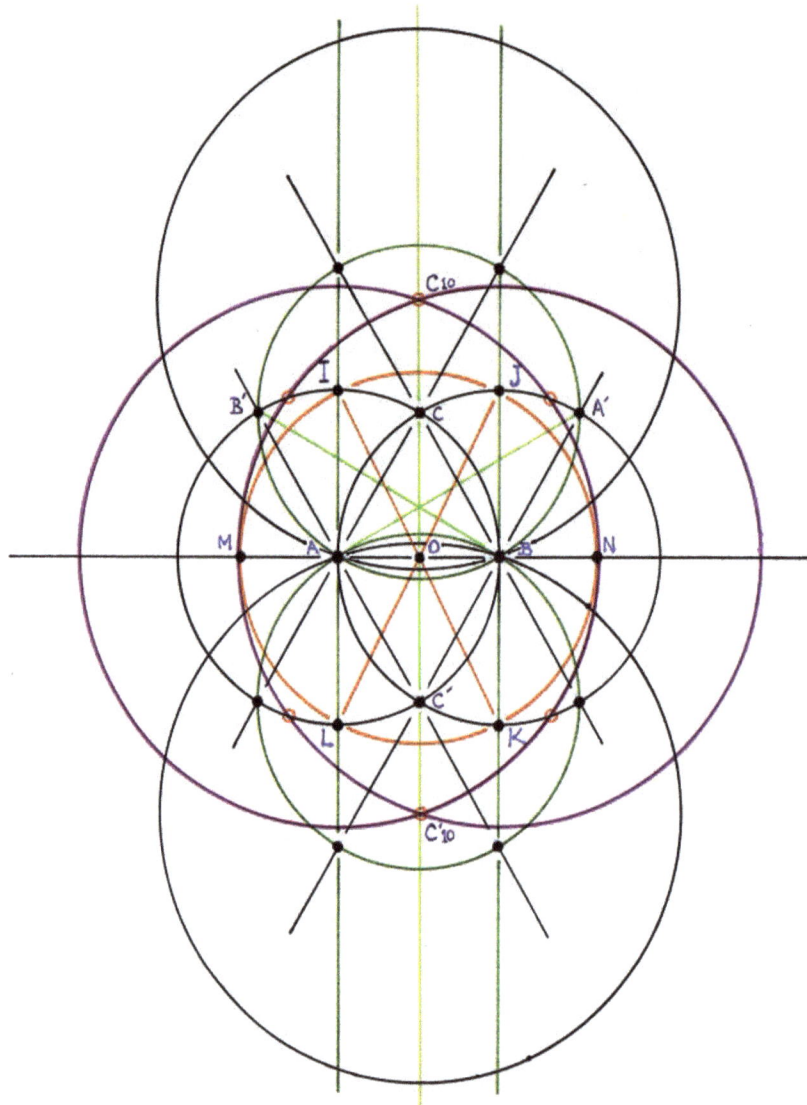

Figure 15. Black circle-pair, radii = Φ

C. The black circle-pair in Figure 15, with radius ½[JL + AB], was derived as an elaboration upon the platform pattern, I, J, K, & L, the same platform that permitted construction of the blue circle-pair.

The blue circle-pair in Figure 12, with radius ½[AJ], was derived as an elaboration upon the platform pattern, the four black triangle lines in Figure 7, 'complements' of (orthogonals to) the platform AA' & BB', which permitted construction of the green circle-pair.

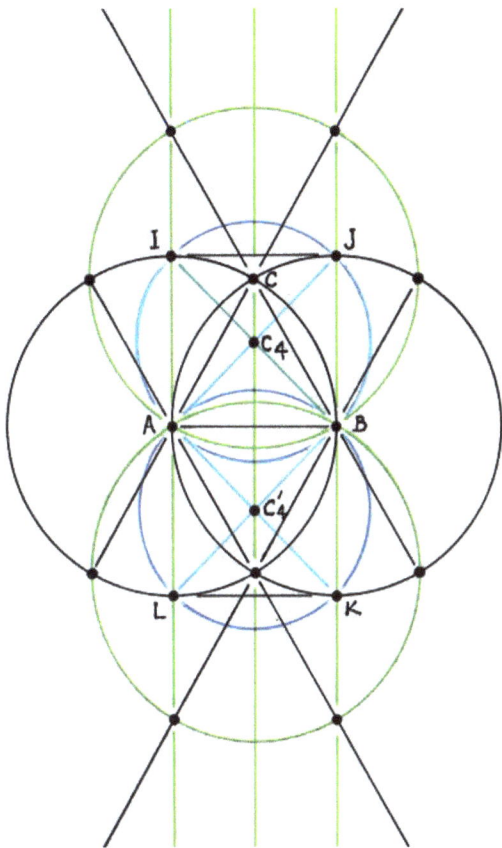

Figure 12.

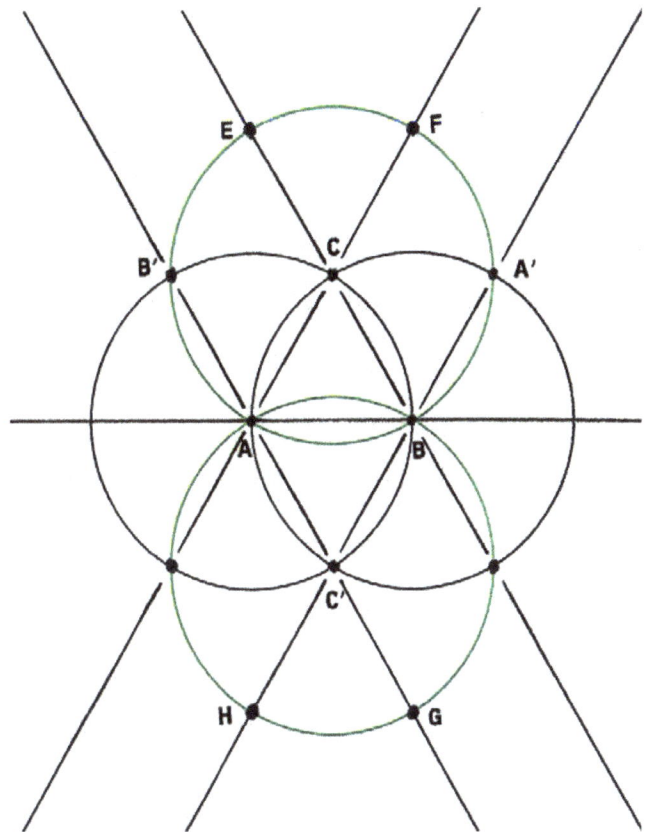

Figure 7.

The green circle-pair in Figure 11, with radius (⅓)AA′, was derived from the implications, Figure 4, augmented by the green circle-pair centered at CC′.

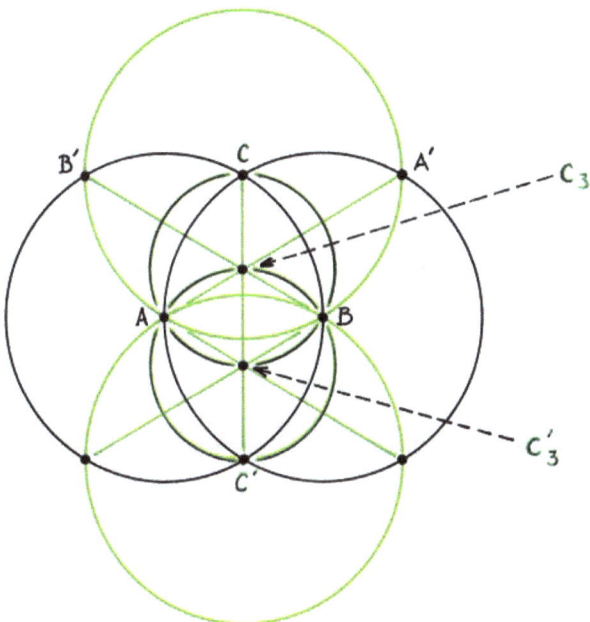

Figure 11.

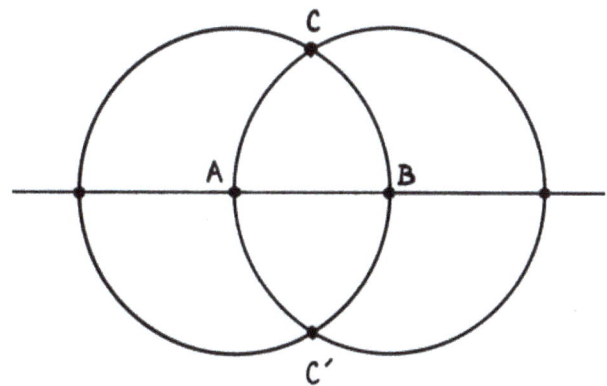

Figure 4.

This hierarchy of circle-pair constructions, with centers at C_3 & C'_3, C_4 & C'_4, and C_{10} & C'_{10}, preserve the reference length as a shared chord. The shared chord no longer makes a *regular* triangle-pair with the hierarchy of radii, (⅓) [AA'], ½ [AJ], and ½[JL + AB].

The space-filling calibration of the Euclidean plane in terms of positional notation of ± powers of the green triangle-pair and the black triangle pair in Figure 5 is shown constructed in Figure 22.

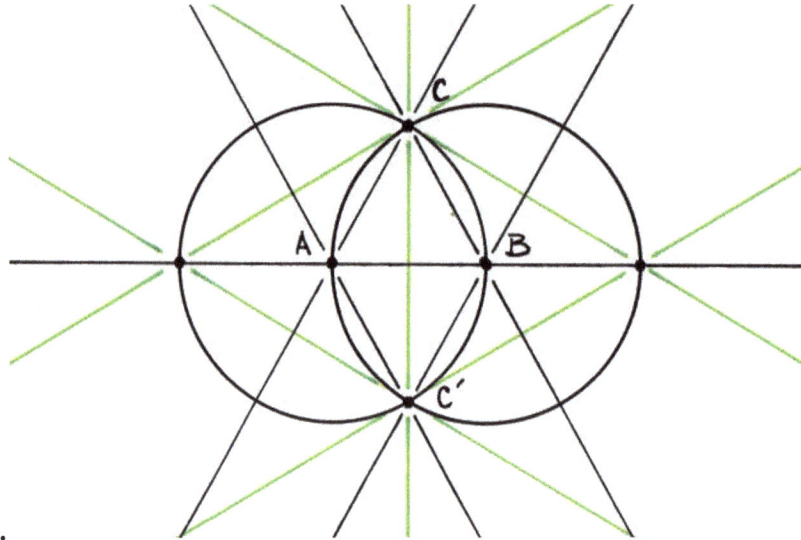

Figure 5.

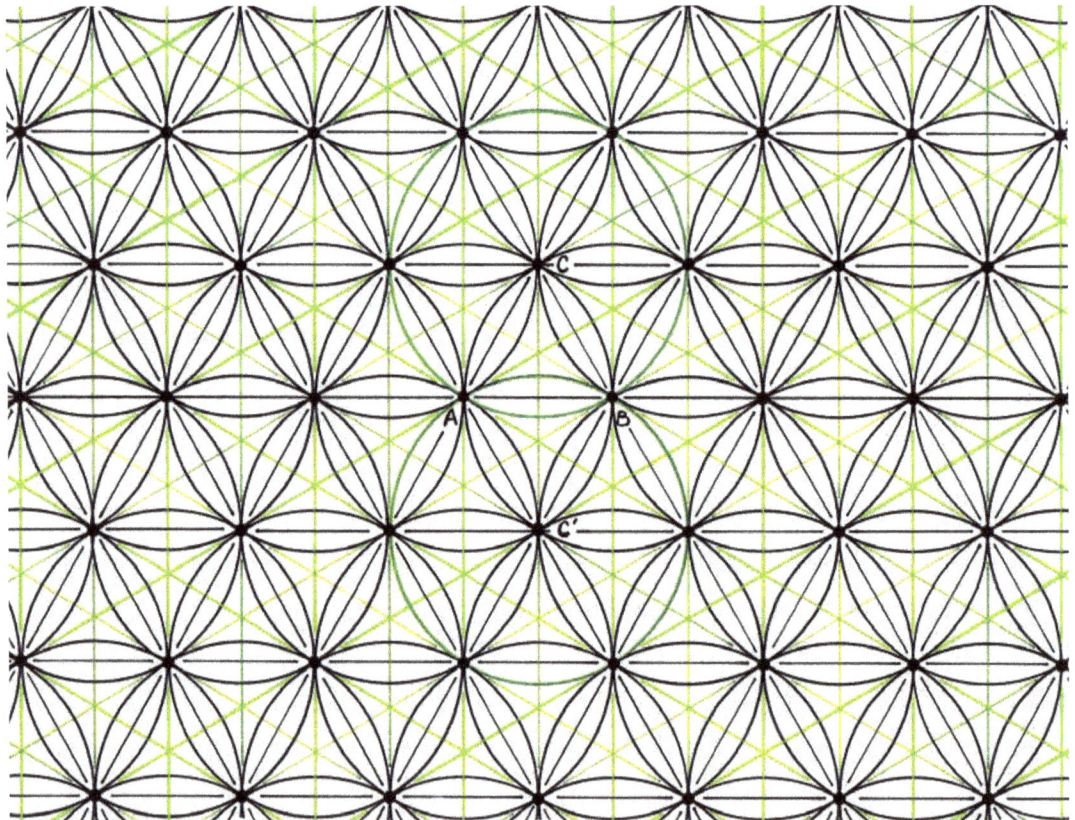

Figure 22. Three ⊥ pairs of √3 Spatial Frequency orientations, the fundamental period.

The space-filling calibration of the plane in terms of positional notation of ± powers of the fractal integer number [AJ] is shown constructed in Figure 26.

The space-filling calibration of the plane in terms of positional notation of powers of the fractal number-pair ½[JL ± AB] aka (½√5 + √1) is shown developed in Figure 27 and constructed in Figure 28a & 28b.

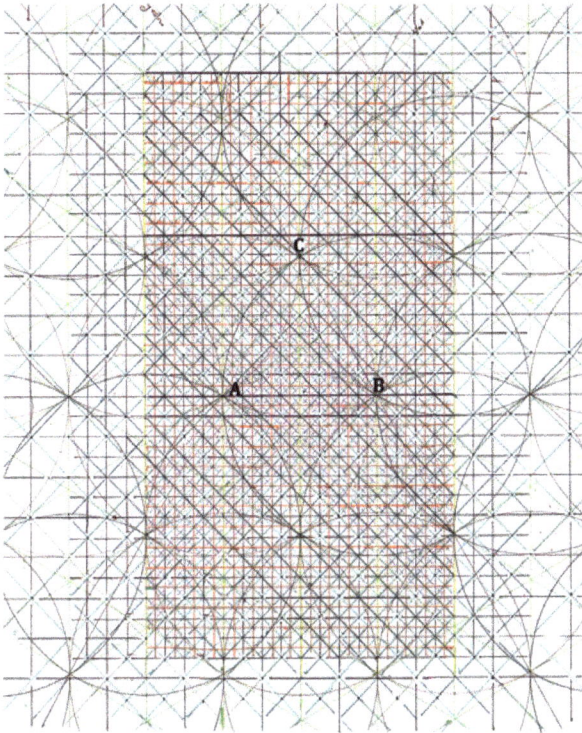

Figure 26. The *planar* space-filling √2 number fractal co-ordinate system. Original vintage drawing by Robert L. Powell, Sr.

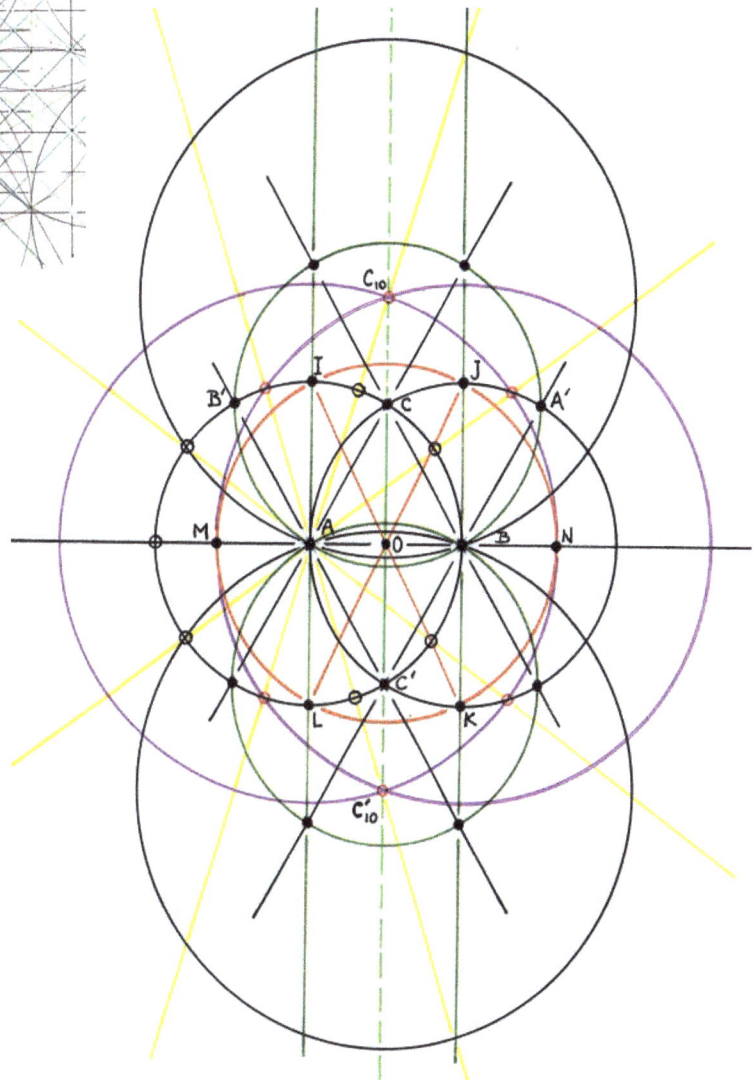

Figure 27. The principal ½[√5±√1] lines.

Figure 28a. Original vintage drawing by Robert L. Powell, Sr.

D. The symmetry-broken *isosceles* triangle-pairs, subtended by the reference element chord at the three-, four- and ten- circle-pair centers, constitute the three fractal number eigen-value selections which calibrate the entire Euclidean plane, with fractal number integer positional notation. (See Figure 27 as framework for Figure 28a & 28b.)

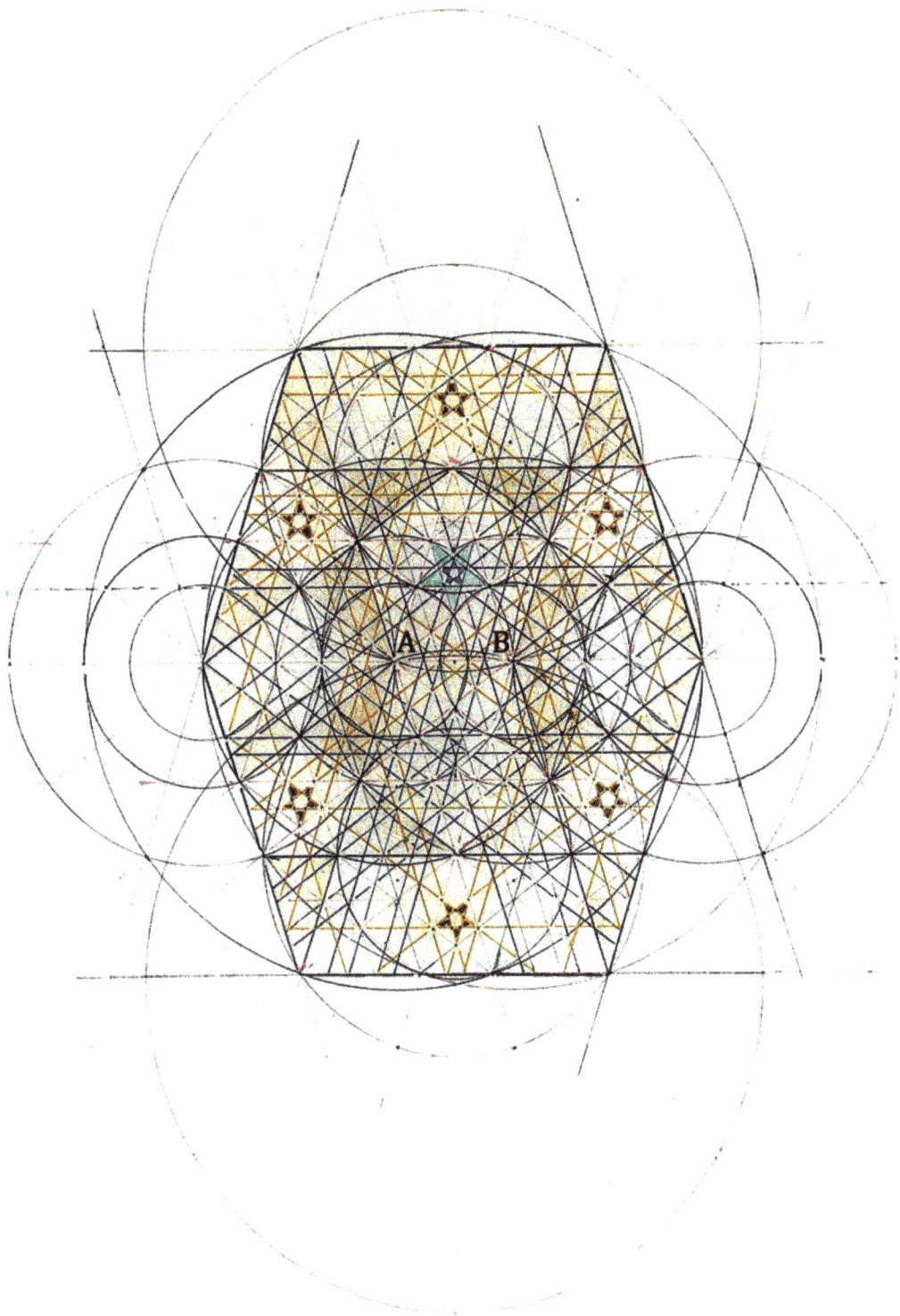

Figure 28b. The planar space-filling ½[√5±√1] fractal number1 co-ordinate system. Original vintage drawing by Robert L. Powell, Sr.

This patterned lattice (Figure 28a & 28b) of space-filling fractal number position-vector coordinate calibrations, is derived as implications embodied in Proposition 1, Book I, of *The Elements*.

Chapter III.   A PLANAR QUALITATIVE PROPERTY OF NUMBER

A. In Chapters I & II we have exposed a here-to-fore *un*-known universe of *quantitative* relations among the *two*-dimensional elements of a countably infinite *planar* set that is also ordinal and organized in reciprocal pairs relative to some element *selected* as the standard, the reference, the norm.

This *planar* and *axial vector* set of elements conduces to the mathematical definition of a positional notation Fractal Number calibration of the Euclidean *plane*—a calibration which subsumes the (2nd millennium's) 'rational' number calibrations of one of a Euclidean plane's *un*-countably many *one*-dimensional Number *lines*.

1. The *standard*, the *reference*, the *norm* selected to enable this epistemological uncovering was a planar **vector** reference element determined by the Humanware's arbitrary positioning of a binary point pair in the plane.

The exposure follows from strategic use of the *complete* set of implications for such an otherwise *un*-marked 'Euclidean' **plane**, embodied in the application of three of Euclid's five Postulates[7] to any **architecture** that is *constructed* on the foundation of a **Humanware's** arbitrary choice of location for the **binary point-pair** reference.

Figure 4 shows the *complete* set of implications of the reference binary point-pair as a *calibration* of, an *initialization* for, the otherwise virgin location for coherent *point-patterns*.

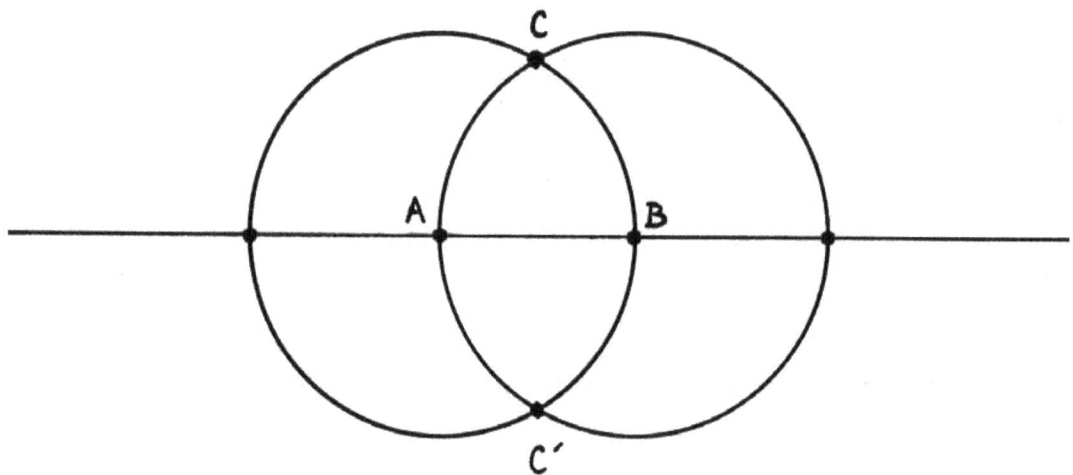

Figure 4.

Of this set of *five* implications, Chapters I & II have mainly exploited the coherent point-pattern producing potential of a *pair*:

the circle-pair that shares the radius determined by the reference *binary*.   [9]

The complementary pair of theorems developed in Chapters I & II can be said to emerge as the reward to a student who explores *further* the Teachings implicit in the geometric form *complex*, [9], as guided by the Teachings educed by Proposition 1, Book I of Euclid's *The Elements*.

2. Although rich in *quantitative* 'new discovery' about the mathematical notion 'Number', the complementary theorem-pair of Chapters I & II *also* teases the Humanware with hints at a *qualitative* property of the *circle* that is **the** *quantitative* connection, in mathematics, between the hierarchy of Number class: Transcendental Number integer; Fractal Number integer; and our mis-named 'Natural' number integer.

3. The symmetry relationship, [10], between

    a circle, the quintessential representation of a Transcendental number,
    and *any* one of its *in*-countably many diameters,   [10]

    embodies **the** *quantitative* ratio, for mathematics, connecting the hierarchy of Number classes: the Transcendental Number integer, the Fractal Number integer, and our mis-named 'Natural' Number integer.

    In this chapter, we study the consequence of a strategic *non*-canonical *breaking* of the symmetry relation established by the *complex*, [10].

B. Figure 16a shows the complex, a circle and any one of its countlessly many radii. Figure 16b shows canonical extensions of the radius, producing its associated diameter. The diameter-circumference complex expresses the symmetry relation, [10]. This is the two-fold symmetry relation we proceed to *break*.

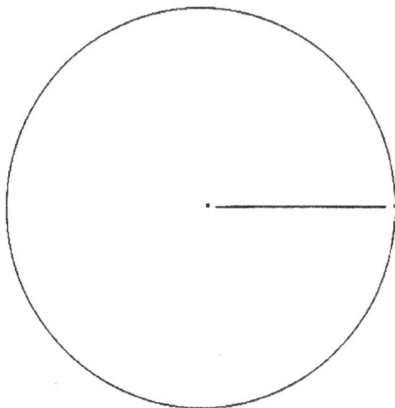

Figure 16a. A circle, and one of its *many* radii

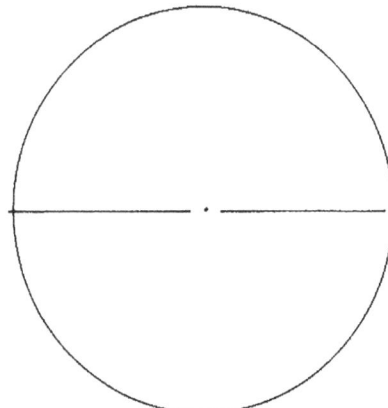

Figure 16b. circle, radius extended to diameter

Figure 17a shows the Humanware-introduced *non*-canonical point, C, located arbitrarily on the circumference. The two-fold symmetry of the figure is thereby broken.

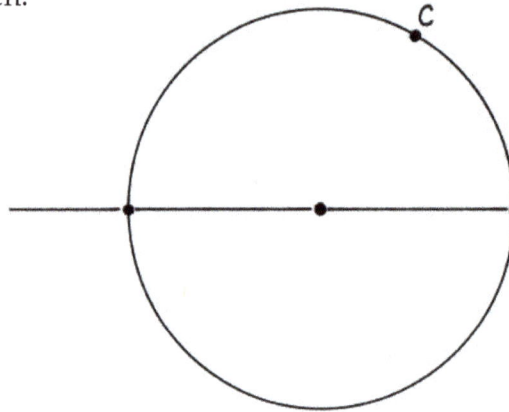

Figure 17a. Non-canonical point, introduced on circumference

This symmetry-breaking point, C, determines a line segment-*pair*: the red line-blue line *complex* shown in Figure 17b.

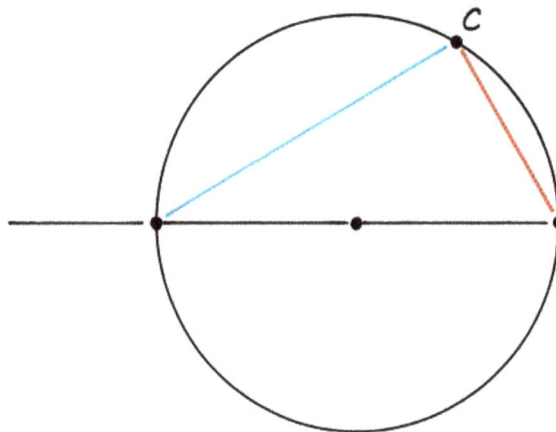

Figure 17b. The determined red line-blue line segement-pair *complex*

The point, C, *also* determines a Triangle *inscribed* in the *semi*-circle.

This inscribed triangle *provides, embodies* and *preserves* **the only** precise, exact, *invariant* quantitative ratio between the Transcendental Number integer classes {{the definitions [6] and [7]}}and the Fractal Number integer class {{the relations [8]}} defined in Chapter I:

$$\text{Circumference} = \pi \times \text{Diameter} \qquad [11]$$

C. Moreover, this semi-circle inscribed triangle provides and embodies the *necessary* and the *sufficient* definition for a unique *invariant* qualitative symmetry relation between the red line-blue line segment pair as represented in Figure 17c, namely:

Extensions of the segment-pair, at C, *invariably* divide the entire plane in *quarters*.

[12]

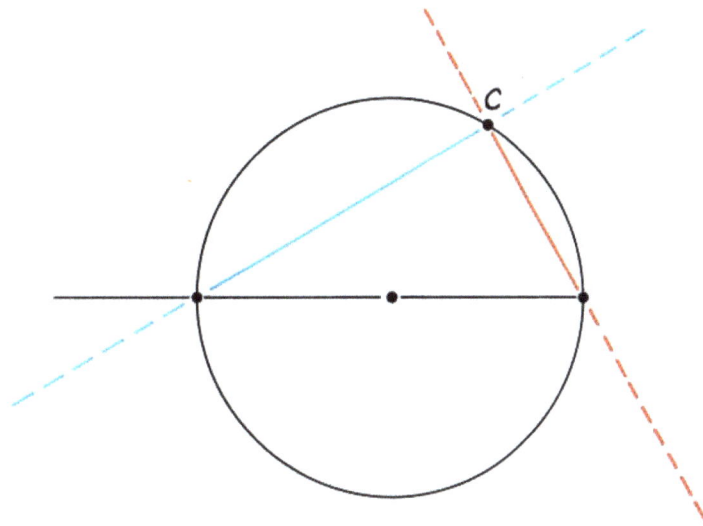

Figure 17c. The determined red line-blue line segment-pair *complex*, with pair extensions

This quadrature division of the plane introduces into the Theory of Number the invariant qualitative symmetry relation, *planar perpendicularity*, as a potential property of *pairs* of Fractal Numbers.

A *corollary* of this 'Right Angles' Postulate[8] is the so-called Pythagorean 'theorem'. A sketch of this corollary is indicated in Figure 18a & Figure 18b.

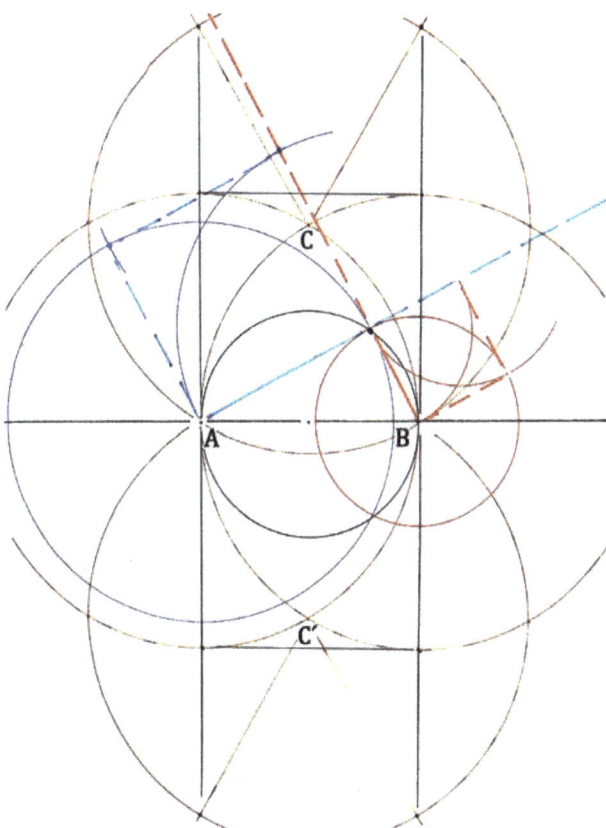

Figure 18a. Sketch of the quadratic law of addition for fractal integer numbers

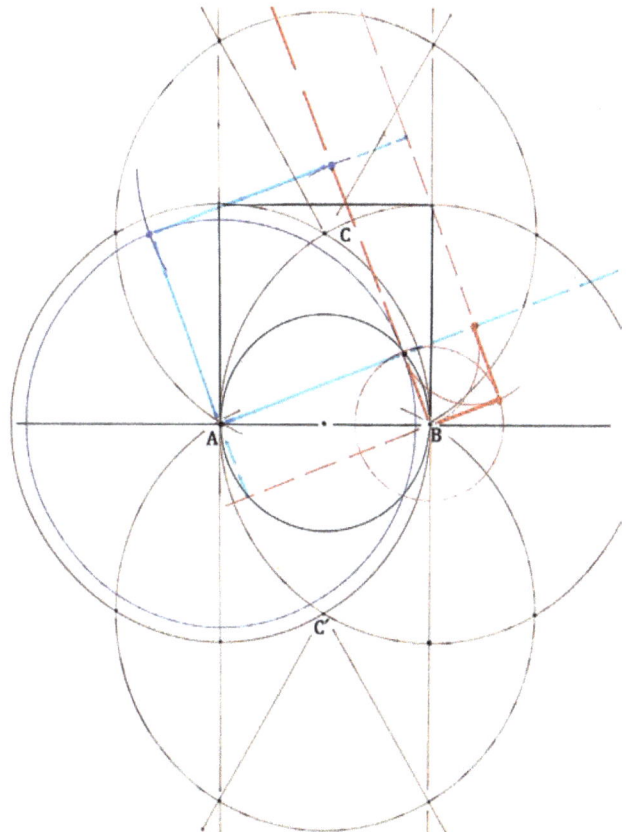

Figure 18b. Sketch of the 'Pythagorean' Triangle Law of quadratic addition.

The first occurrence of this quadrature division in Euclid's *The Elements* is permitted [as an exercise for the student's Humanware] in Proposition 1, Book I. There, given the reference point-pair, the student is taught to construct the circle-pair, [9].

The point-pair, C & C′ together with reference point binary (AB), produces the mutually perpendicular [and mutually bisecting] elemental implication of the Laws of Construction. (See Figure 4.)

The quadrature division property of the 'Right Angle' applies to the three green line-black line complexes in Figure 5. These three perpendicular line-pair complexes validate the provisional assumption of the Pythagorean Theorem in the Chapter I derivations.

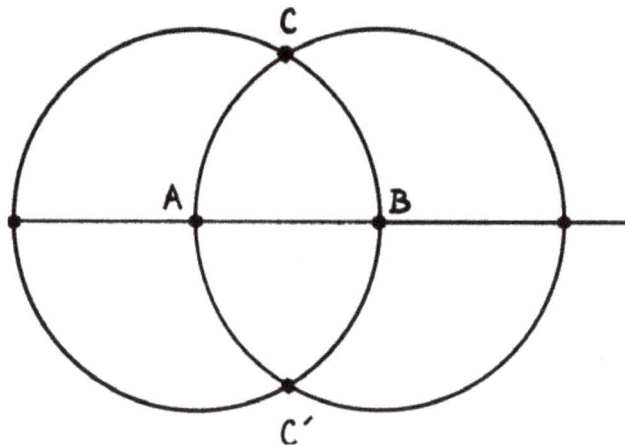

Figure 4.

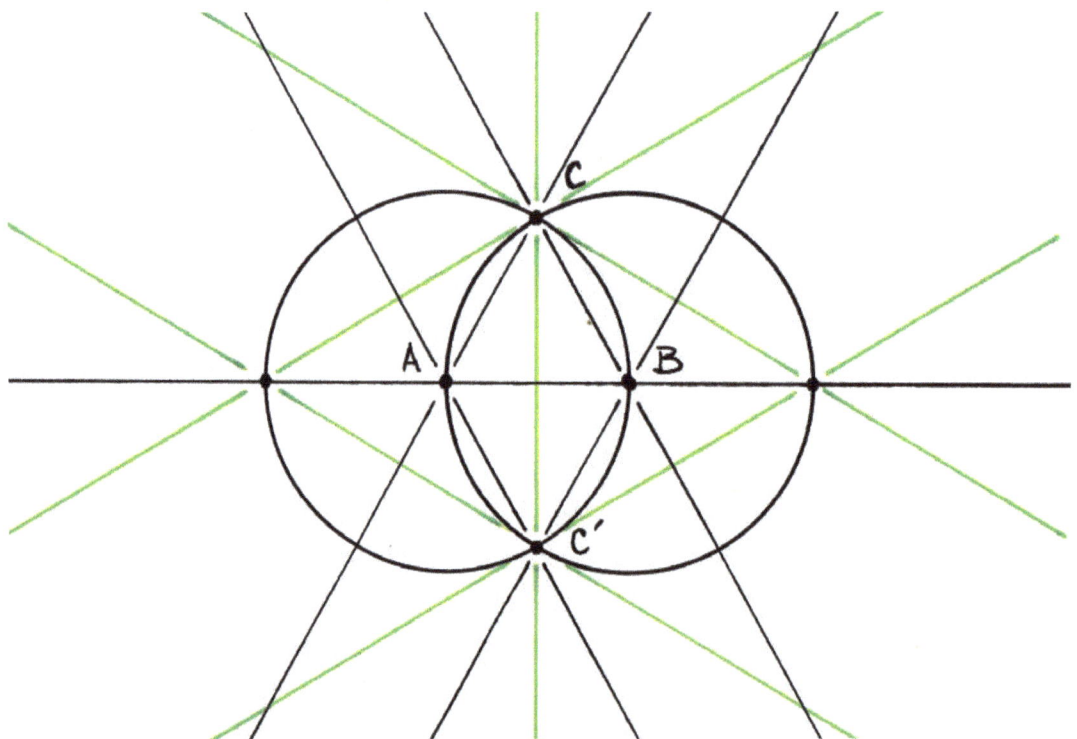

Figure 5.

7 David Bergamini, et al [Rene Dubos, Henry Margenau, C. P. Snow], eds. Life Science Library *Mathematics*. New York: *Time Incorporated*. 1963. p.46.

8 Ibid.

Chapter IV   DERIVATION OF A FRACTAL INTEGER NUMBER TRIGONOMETRY TABLE: A Revisit of the Theorem Derived in Chapter I

A. The fractal number integers *defined* in the *canonical complex*, Figure 10b of Chapter I, can be interpreted as the **_computed_** values of the quadrature set of radii for each transcendental number integer circle-pair.

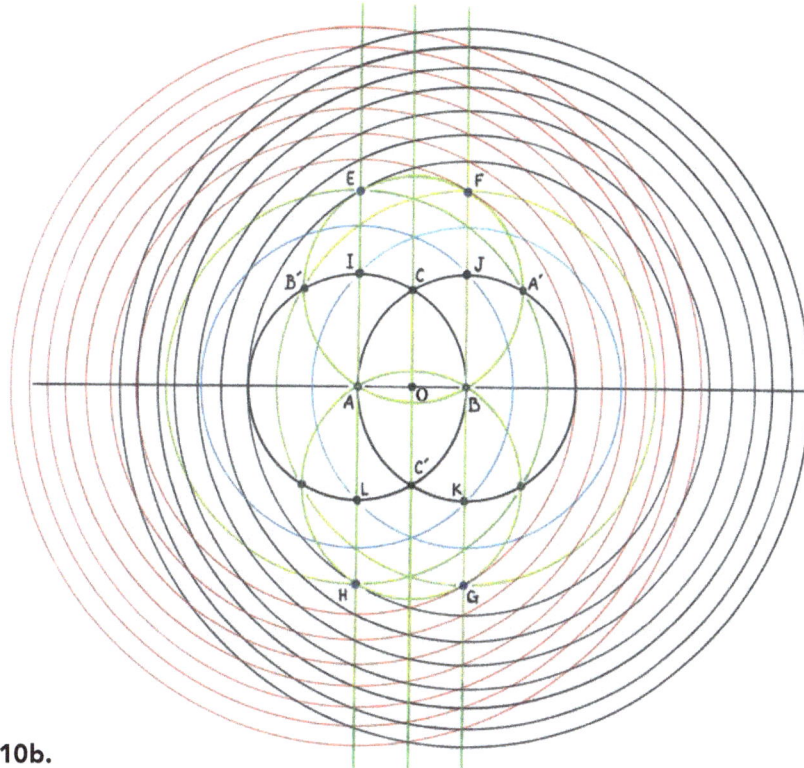

Figure 10b.

The computed values may also be interpreted as a sub-set of the *solutions* to the system of 'simultaneous' algebra equations which represent the 'parallel-organized' geometry *complex*, Figure 10b:

the one black line;
the three articulated green lines;
the (n) articulated circle-pairs, where n = 1, 2, 3, … .   [13]

An algebra equation system representation of the parallel-organized *computation*, [13], is

$$y = 0;$$
$$x + \tfrac{1}{2}AB = 0;$$
$$x = 0;$$
$$x - \tfrac{1}{2}AB = 0;$$
$$[x_m + \tfrac{1}{2}AB]2 + [y_m]2 = [r_m]2, \qquad m = 1, 2, 3, … ;$$
$$[x_n - \tfrac{1}{2}AB]2 + [y_n]2 = [r_n]2, \qquad n = 1, 2, 3, … .$$ [14]

This <u>sub-set</u> of the circle-circle solutions has been contrived to locate *on the green lines*, a consequence of the <u>series</u>-organized computation program strategy employed to construct [13].

B. A casual inspection of Figure 10b invites the question:

What about *the <u>rest</u>* of the *circle-circle* system's solutions indicated in the Figure 10b?

To 'discover' and expose the <u>complete</u> set of *fractal number* <u>solutions</u> for the geometric articulation of *transcendental number integer* family-pair, Figure 10b, is the explication task of this chapter.

HIATUS:

"The originality of mathematics consists in the fact that in mathematical science connections between things are exhibited which, apart from the agency of human reason, are extremely unobvious. Thus the ideas, now in the minds of contemporary mathematicians, lie very remote from any notions which can be immediately derived by perception through the senses; unless indeed it be perception stimulated and guided by antecedent mathematical knowledge."

– A. N. Whitehead, *Science and the Modern World*, the Macmillan Co. (1925); Mentor Books (1948, 1950), chapter 2, p. 19; *The Free Press* (1953) p. 25.

C. The *complete* set of circle-circle solutions is discovered to compose the not-immediately obvious organized system of scaled *fractal number* replications of the *complex* of *fractal number integers* constructed by the right triangle *pairs* that are parallel-organized in the structure, Figure 10b.

The period determined for the cycle of scaled replications is established by the vector AB, the separation of the concentric family pair centers.

The vector AB enables canonical construction of the plane-filling pattern of shared-radius circle-pairs shown in Figure 19a. Figure 19a in turn enables the plane-filling pattern of periodic replications of the three green line pattern in Figure 10b. This pattern of green line replications is shown in Figure 19b.

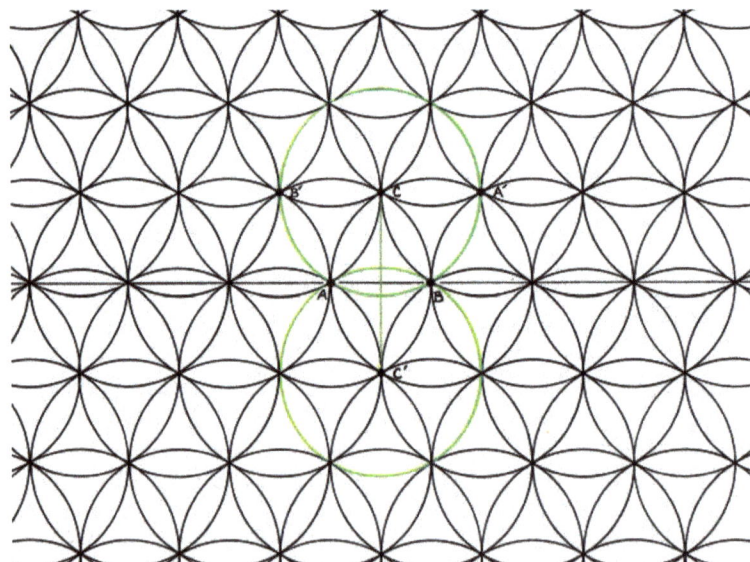

Figure 19a. A plane-filling replication of the shared radius

D. 1. The vector articulation, AB, of the circle family pair centers is such as to organize all the circle-crossings into *pair-wise* co-locations, only on the green lines of Figure 19b.

2. Moreover, the pattern of co-location of pair-wise solutions on the green line cycles of Figure 19b *preserves* the pattern of pair-wise solution locations on the three green lines of Figure 10b.

3. The unexpected invariance of the pattern of pair-wise solution locations on the countably infinite family of green lines is emphasized in Figures 19c-19d.

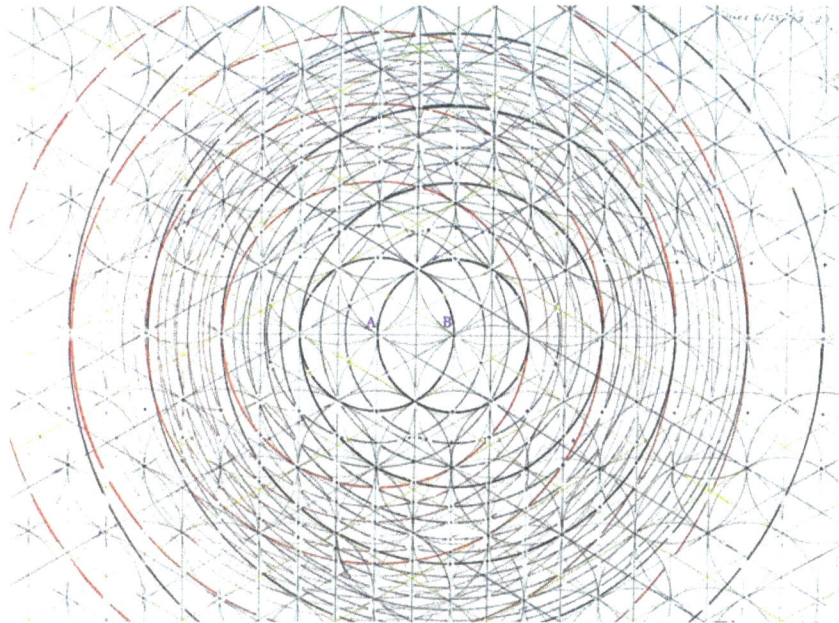

Figure 19c. Original vintage drawing by Robert L. Powell, Sr.

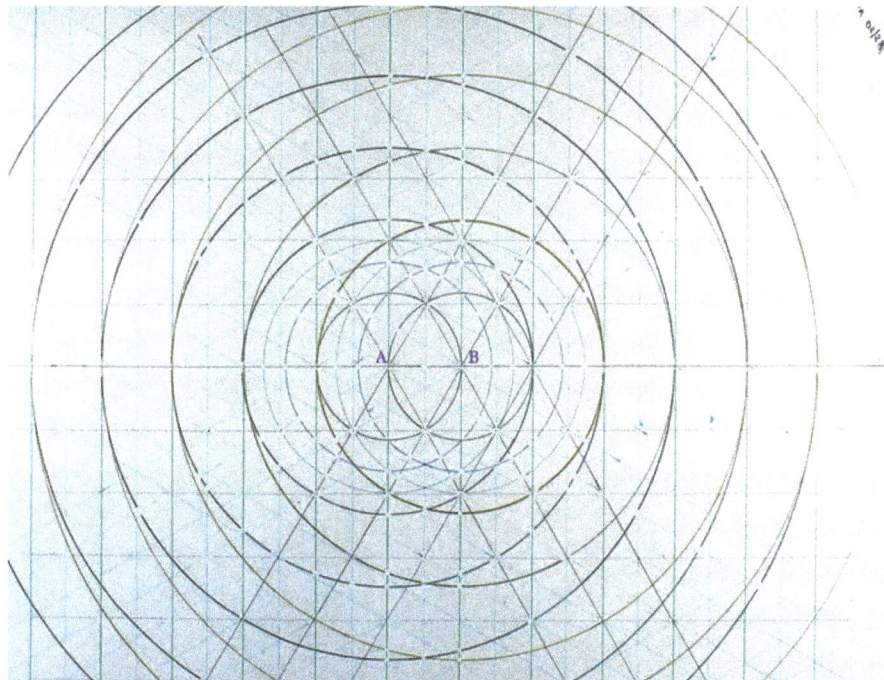

Figure 19d. Original vintage drawing by Robert L. Powell, Sr.

E. With the A-point as reference, *pattern* of the pair-wise solution location *computations* can be tabulated. With the center of the reference circle at the A point as origin, the 'x' component of the Cartesian coordinate of *each* pair-wise solution location sub-set is quantized to some number of units of ½[AB].

The 'y' component of each pair-wise solution location is determined by the particular radius of *the* 'A-centered' circle providing the particular solution.

An equivalent restatement of the above description of the coordinates of the circle-circle solution location *computations* is:

Each solution location determines a right triangle with hypotenuse being the radius of a transcendental number circle centered at the point A.

The 'x' component of the right triangle is some multiple of the length interval, ½AB.

The 'y' component of the right triangle side is thus determined as some length interval along a 'green line'.

F. Using this equivalent scheme, the *fractal number computations* can be tabulated for the *complete* set of solutions for the parallel-organized articulated *pair* of countably infinite *transcendental number integer* families represented in Figure 10b.

Figures 19c and 19d show four quadrants of symmetry about the midpoint of AB.

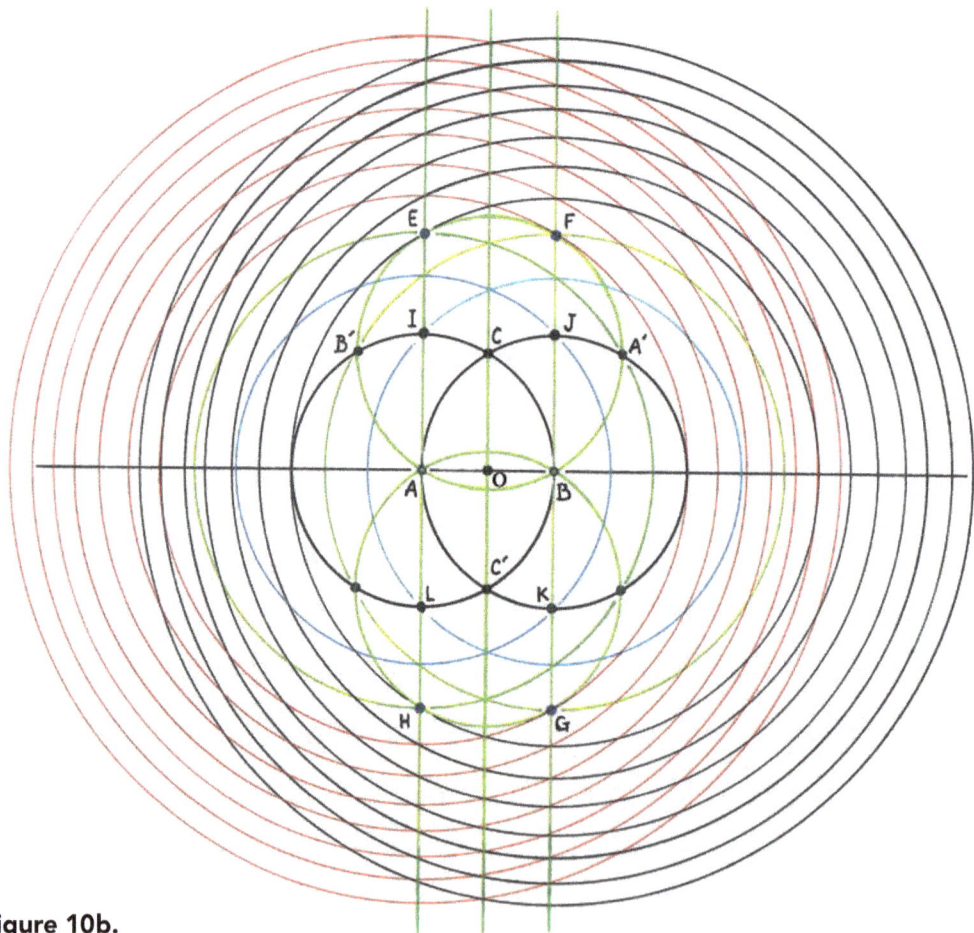

Figure 10b.

Appendix III (a table of Fractal Number Trigonometric Ratios, justified as a Theorem of Euclid.) gives the fractal number tabulations for the 'first' quadrant of Figure 10b solutions, using the 'equivalent' scheme:

A table of right triangles, with
$$[\text{hypotenuse}]^2 = [\sqrt{n}]^2 = n$$
$$\text{'x' side} = 2m\ (\tfrac{1}{2}\sqrt{1})$$
$$= [2m+1]\ (\tfrac{1}{2}\sqrt{1})$$
$$\text{'y' side} = [\sqrt{n}]^2 - ['x']^2 \qquad\qquad [15]$$

G. The tabulation, Appendix III, constitutes a table of Fractal Number Trigonometric Ratios, justified as a Theorem of Euclid.

H. If the 'beginning'-and-'end' of the 'reference' cycle are taken as at the reference point-pair, then the reference mid-cycle is 'at' the mid-point, O.

1. In Figure 10b let us consider the family of circles, centered at the point A, with radii = $\sqrt{1}$, $\sqrt{2}$, $\sqrt{3}$, ... $\sqrt{n}$, ... With this family, let us ask the question-pair, (1) what is the intersection of the nth circle with the extension of line CC' and (2) what is the intersection of the nth circle with the green line thru B?

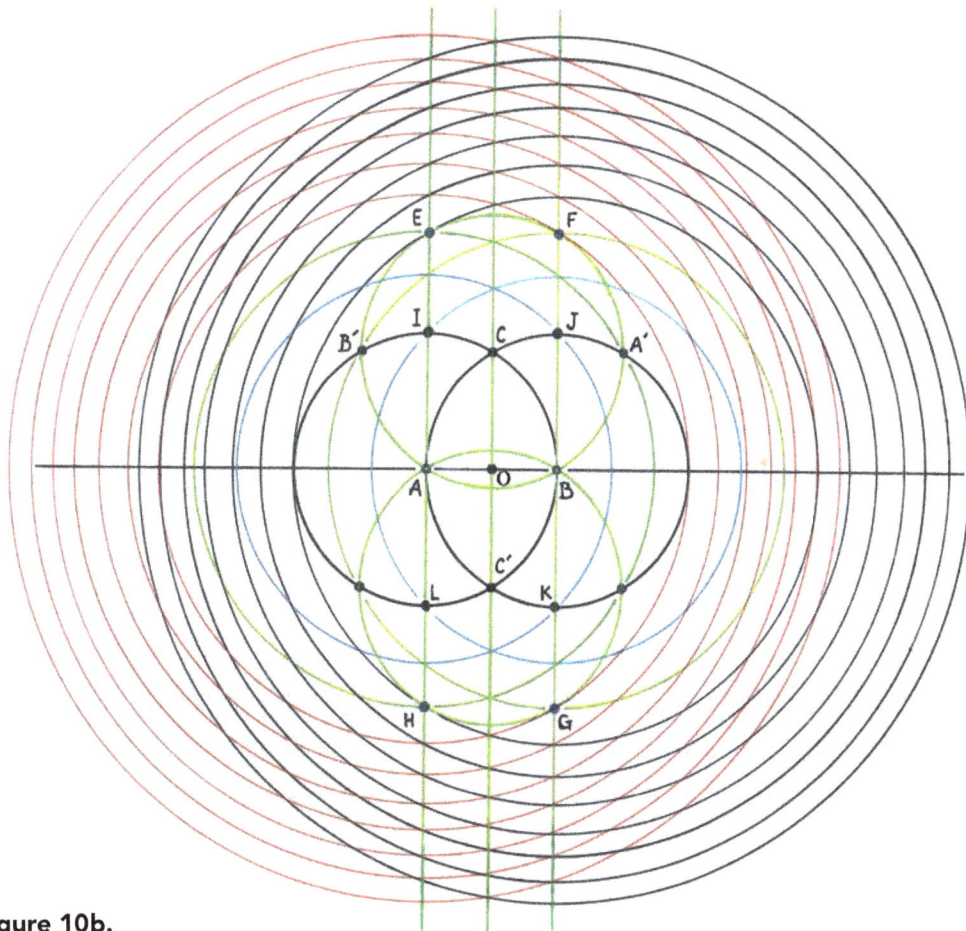

Figure 10b.

Since both green lines are ⊥ to AB, each of the two sets of intersections produces a family of right triangles. In one family the hypotenuse set is √1, √2, √3, ..., √n, ...; one side of the set of right triangles is *fixed*, at ½ AB [where ½AB = ½(√1), = (√1)/2].

For the second family of intersections the hypotenuse set is √2, √3, √4, ..., √n, ...; one side of this set of right triangles is *fixed*, at (2/2) AB = 2(√1)/2.

2. This pair of relations determines the series of the *third* side for each right triangle series:

hypotenuse	1st series fixed side	1st series, 3rd side	2nd series fixed side	2nd series, 3rd side
√1	(√1)/2	√[1-(¼)] = (√3)/2		
√2	"	√[2-(¼)] = (√7)/2	2(√1)/2	√[2-1] = √1
√3	"	√[3-(¼)] = (√11)/2	"	√[3-1] = √2
√4	"	√[4-(¼] = (√15)/2	"	√[4-1] = √3
√5	"	√[5-(¼)] = (√19)/2	"	√[5-1] = √4
√6	"	√[6-(¼)] = (√23)/2	"	√[6-1] = √5
√7	"	√[7-(¼)] = (√27)/2	"	√[7-1] = √6
√8	"	√[8-(¼)] = (√31)/2	"	√[8-1] = √7
√9	"	√[9-(¼)] = (√35)/2	"	√[9-1] = √8
√10	"	√[10-(¼)] = (√39)/2	"	√[10-1] = √9
...	"	...	"	...
...	"	...	"	...
√n	"	√[n-(¼)]-½√[4n-1]	"	√[n-1]
...	"	...	"	...
...	"	...	"	...

3. A close *visual* inspection of the *geometrical* pattern of the *complete* set of algebraic solutions for the circle-pairs in the [13] of Figure 10b guides the Geometer's Humanware to examine the pair of intersection sets for the circle family centered at the point A, with the next distant cycle of green lines in Figure 19b, namely (a) the green line a distance 3(√1)/2 from CC′ and (b) the green line 4(√1)/2 from CC′.

For the green line with the 3(√1)/2 separation from CC′, the right triangle hypotenuse set is √3, √4, √5, √6, ..., √n, ... as read from Figure 10b.

For the green line with the 4(√1)/2 separation from CC′, the right triangle hypotenuse set, read from Figure 10b, is √5, √6, √7, √8, ..., √n, ...

This pair of (hypotenuse-side) relations determines the series of the *third* side for each right triangle series:

hypotenuse	3rd fixed side	3rd series, *third* side	4th fixed side	4th series, *third* side
$\sqrt{3}$	$3(\sqrt{1})/2$	$\sqrt{[3-(9/4)]} = \sqrt{3/4} = (\sqrt{3})/2$		
$\sqrt{4}$	"	$\sqrt{[4-(9/4)]} = \sqrt{(7/4)} = (\sqrt{7})/2$		
$\sqrt{5}$	"	$\sqrt{[5-(9/4)]} = \sqrt{(11/4)} = (\sqrt{11})/2$	$4(\sqrt{1})/2$	$\sqrt{[5-4]} = \sqrt{1}$
$\sqrt{6}$	"	$\sqrt{[6-(9/4)]} = \sqrt{(15/4)} = (\sqrt{15})/2$	"	$\sqrt{[6-4]} = \sqrt{2}$
$\sqrt{7}$	"	$\sqrt{[7-(9/4)]} = \sqrt{(19/4)} = (\sqrt{19})/2$	"	$\sqrt{[7-4]} = \sqrt{3}$
$\sqrt{8}$	"	$\sqrt{[8-(9/4)]} = \sqrt{(23/4)} = (\sqrt{23})/2$	"	$\sqrt{[8-4]} = \sqrt{4}$
$\sqrt{9}$	"	$\sqrt{[9-(9/4)]} = \sqrt{(27/4)} = (\sqrt{27})/2$	"	$\sqrt{[9-4]} = \sqrt{5}$
$\sqrt{10}$	"	$\sqrt{[10-(9/4)]} = \sqrt{(31/4)} = (\sqrt{31})/2$	"	$\sqrt{[10-4]} = \sqrt{6}$
$\sqrt{11}$	"	$\sqrt{[11-(9/4)]} = \sqrt{(35/4)} = (\sqrt{35})/2$	"	$\sqrt{[11-4]} = \sqrt{7}$
$\sqrt{12}$	"	$\sqrt{[12-(9/4)]} = \sqrt{(39/4)} = (\sqrt{39})/2$	"	$\sqrt{[12-4]} = \sqrt{8}$
$\sqrt{13}$	"	$\sqrt{[13-(9/4)]} = \sqrt{(43/4)} = (\sqrt{43})/2$	"	$\sqrt{[13-4]} = \sqrt{9}$
$\sqrt{14}$	"	$\sqrt{[14-(9/4)]} = \sqrt{(47/4)} = (\sqrt{47})/2$	"	$\sqrt{[14-4]} = \sqrt{10}$
$\sqrt{15}$	"	$\sqrt{[15-(9/4)]} = \sqrt{(51/4)} = (\sqrt{51})/2$	"	$\sqrt{[15-4]} = \sqrt{11}$
...	"		"	
...	"		"	
$\sqrt{n}$	"	$\sqrt{[n-(9/4)]} = (½)\sqrt{[4n-3^2]}$	"	$\sqrt{[n-4]} = (½)\sqrt{[4n-4^2]}$
$\sqrt{[n+1]}$	"		"	
...	"		"	
...	"		"	

4. Comparison of the pair of circle-green line solution sets derived in sub-section D.2., above, with the circle-green line solution sets derived in sub-section D.3., above, shows an interesting correspondence between the solution sets for the selected adjacent cycles of the green-line spatial frequency pattern isolated in Figure 19b.

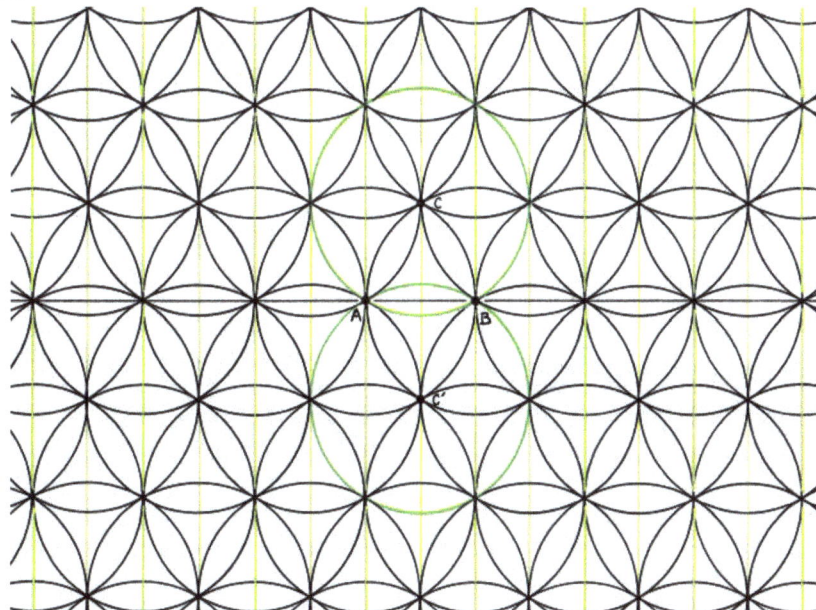

Figure 19b. A plane-filling patterned self-replication of the (green) reference element CC' = √3

For this pair of cycles, the system of circle-green line intersections for the second cycle's mid-line (i.e., at the third fixed distance, $3[\sqrt{1}]/2$, from the A point) *replicates* the system of circle-green line intersections for the first fixed side, $1(\sqrt{1})/2$ {i.e., the CC′ line.}

Similarly, the system of circle-green line intersections on the second cycle's end-line (i.e., at the fourth fixed distance, $4[\sqrt{1}]/2$, from the A point) *replicates* the system of circle-green line intersections for the second fixed distance, $2[\sqrt{1}]/2$, from the A point.

Figure 10b shows that the system of circle-line intersections on the green line at the fixed distance $2[\sqrt{1}]/2$ *replicates* the system of circle-green line intersections on the green line passing thru the A point {i.e., the green line with fixed distance, *zero*, from the A point.}

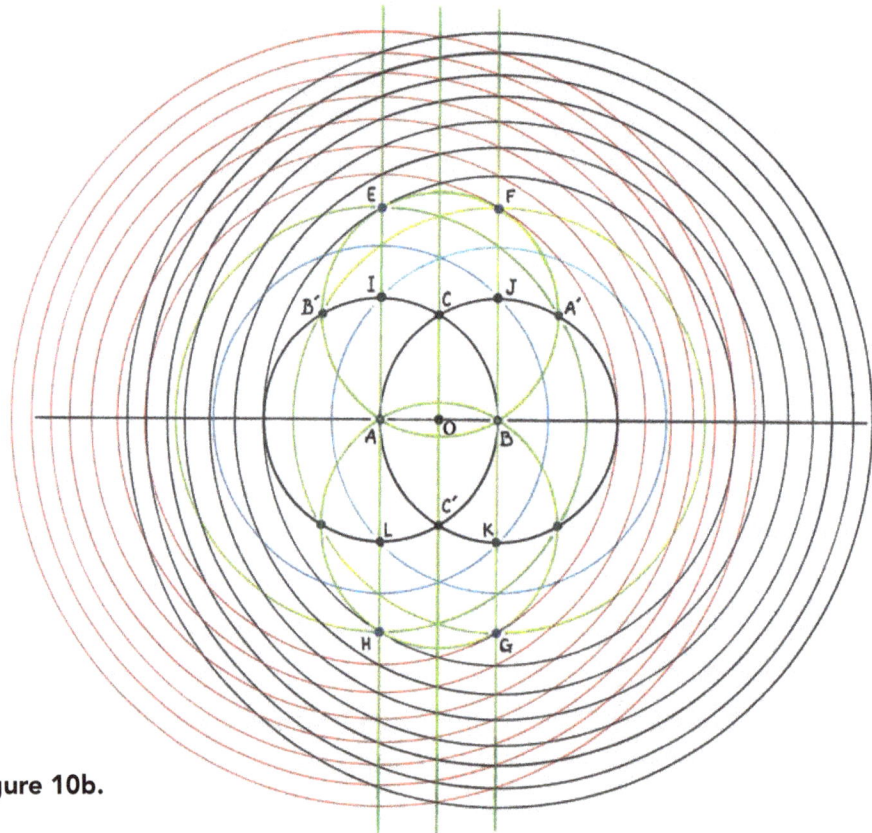

Figure 10b.

5. A careful construction of the concentric circle family-pair of Figure 10b, upon the green-line spatial frequency background of Figure 19b, shows the **_computation_** of the *complete* set of solutions for the system of circle-pairs in [13] above. The Pythagorean Triad of Fractal Integers and their associated Fractal Number ratios, *computed* for a First Quadrant of the entire Euclidean plane, relative to the *vector* fractal integer $AB = \sqrt{1}$, are shown in APPENDIX III.

6. The Pythagorean Triad connecting the Fractal Number Integers, $\sqrt{1}$, $\sqrt{2}$, $\sqrt{3}$, ..., $\sqrt{n}$, $\sqrt{(n+1)}$, ... , and their associated countably infinite Fractal Number ratios tabulated in APPENDIX III constitutes a doubly infinite Table of **Fractal Number Trigonometric ratios.**

Chapter V **DERIVATION OF THE THREE PLANE SPACE-FILLING FRACTAL INTEGER NUMBER ORTHOGONAL COORDINATE SYSTEMS: A Revisit of the Theorems Derived in Chapter II**

In this Chapter we extract, from the Teachings implicit in Euclid's Proposition 1, Book I *still _more_* meta-mathematical implications of *the _circumference_* of a circle.

In particular, we examine the fractal integer number _geometric_ _foundation_ for the EQUATIONS DEFINING SECTIONS OF A CIRCUMFERENCE.[9]

We make a *Fractal Number _geometric_* excursion into *CYCLOTOMY,* the circumference-sectioning Theory discussed *algebraically* by Gauss.[10]

Our study of the Cyclotomy problem is more general than the approach used by Gauss. His dissertation tacitly assumes the rational domain of number sufficiently rich basis within which to express the *variables,* as well as the *coefficients,* of the _equations_ defining the *compass constructible* regular polygon.

Our study disregards his emphasis on Cyclotomy as study of the _algebraic_ theory of the equation for the nth roots of unity. Alternatively, we exploit the _geometric_ foundation, _synthesizing_ the equations for the cyclotomic circumferences _and_ the equations' regular cyclotomic pattern of _solutions_—as canonical developments from the very first Proposition in the very first Book of *The Elements.*

The results of our study emphasized in this Chapter is an introduction of the Fractal Number Theory into Descartes' contribution in *Analytic Geometry*[11]: the fixing of a point's location in a plane by assigning a number-pair, co-ordinates, giving its distance from two mutually perpendicular reference lines.

Our Fractal Number based Analytic Geometry result introduces a canonical *hierarchy* of three co-ordinate number base systems. The most primitive number system basis is

$$\{\sqrt{1}, \sqrt{3}, [\sqrt{3}]^{-1}\}. \tag{16}$$

This number basis system's construction permits construction of the number system basis

$$\{\sqrt{1}, \sqrt{2}, [\sqrt{2}]^{-1}\}. \tag{17}$$

The number system basis, [15], provides the canonical foundation for construction of what is arguably the most *intelligent* number system basis possible in a *mathematics system*:

$$\{\sqrt{1}, \tfrac{1}{2}[\sqrt{5} + \sqrt{1}], \tfrac{1}{2}[\sqrt{5} - \sqrt{1}]\}. \tag{18}$$

We set out now to develop these three implications of Cyclotomy, from Proposition 1, Book I.

A. The planar, √3 basis, fractal number system.

1. The plane-filling pattern of replications of the segment CC′ is shown Figure 19b.

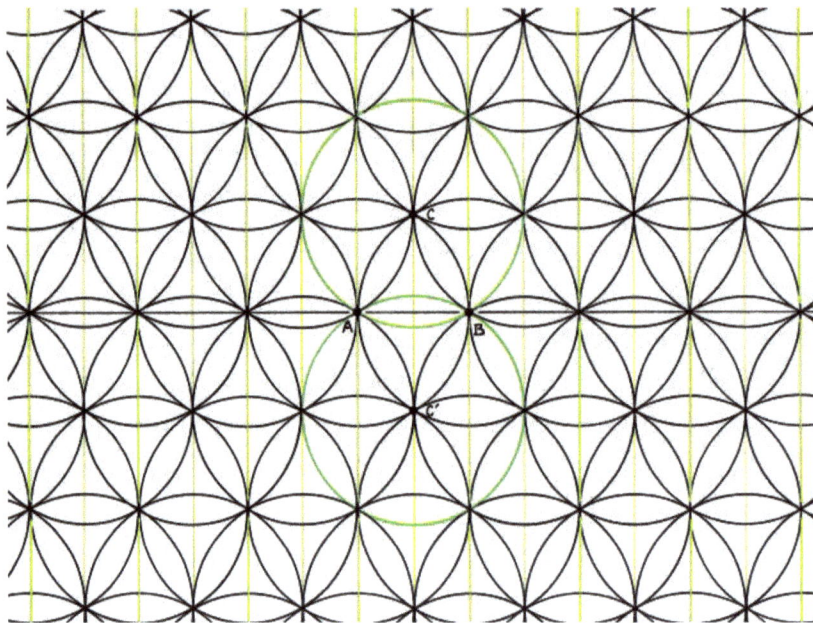

Figure 19b. A plane-filling patterned self-replication of the (green) reference element CC′=√3

It has a pair of symmetry partners: the infinite pattern of replications of AA′ and the infinite patterns of replications of BB′. The generators, AA′ & BB′, of this pair of partners is produced in Figure 20a.

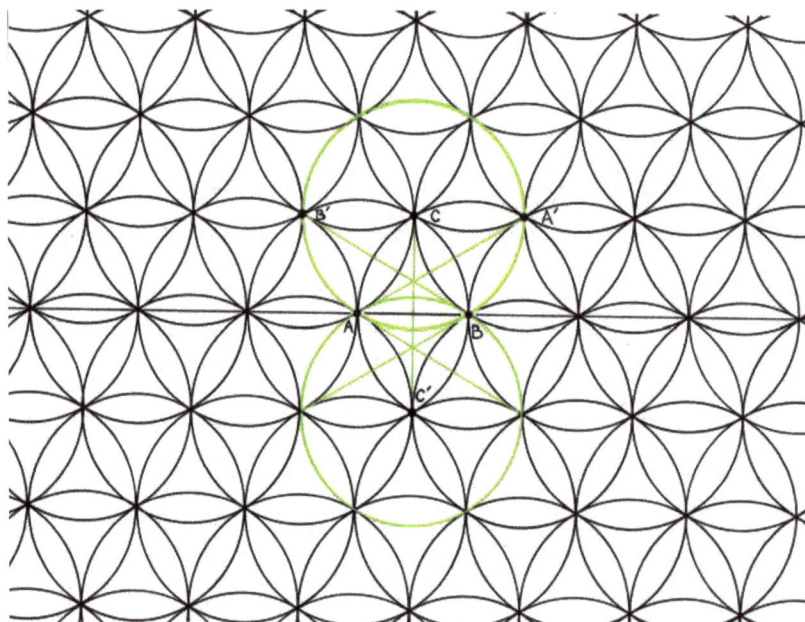

Figure 20a. The √3 Spatial Frequency Generators, AA′, BB′ and CC′

The superposition of this (green) trio of infinite patterns of spatial frequencies is shown in Figure 20b.

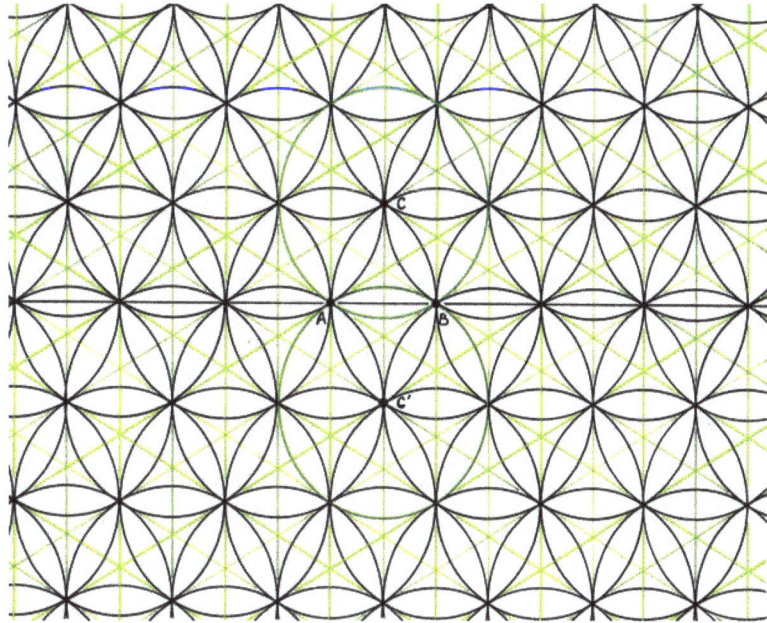

Figure 20b. Three √3 Spatial Frequency orientations, the fundamental period.

Figure 4 permits a (black) triangle-pair, inscribed in the space common to the reference circle-pair, that *shares* the reference segment AB as a common side.

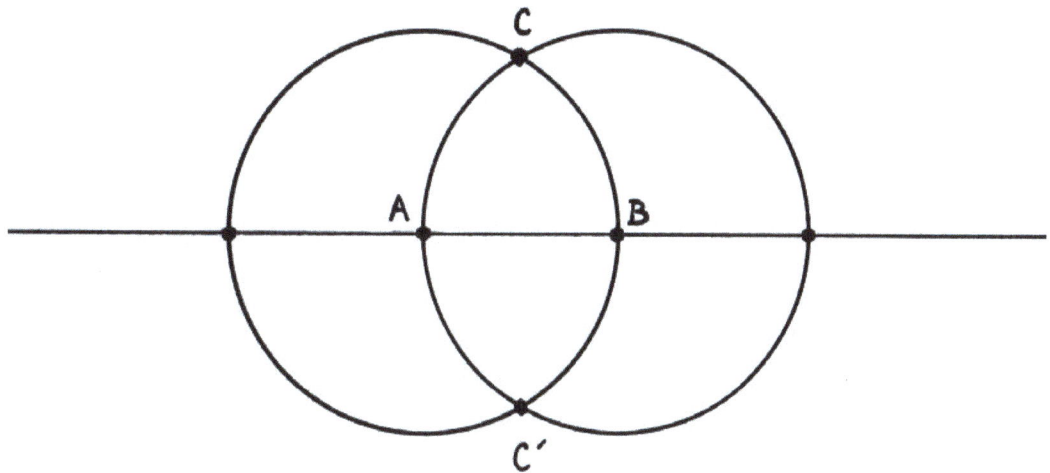

Figure 4.

Figure 21a shows this triangle-pair in the context of the plane-full of circle-pair replications. The circle-pair replications permit a plane-filling pattern of replications of the black triangle-pair.

The result is the superposition of the trio of (black) spatial frequency patterns shown in Figure 21b.

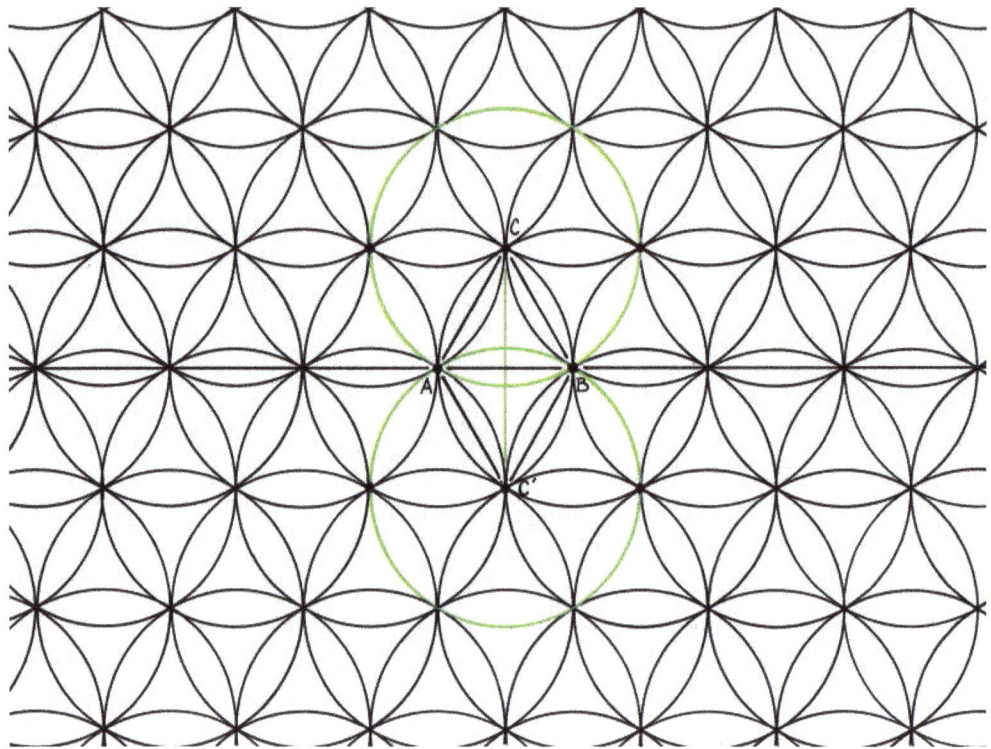

Figure 21a. Three √3 Spatial Frequency Generators, AB; AC & BC'; and AC' & BC

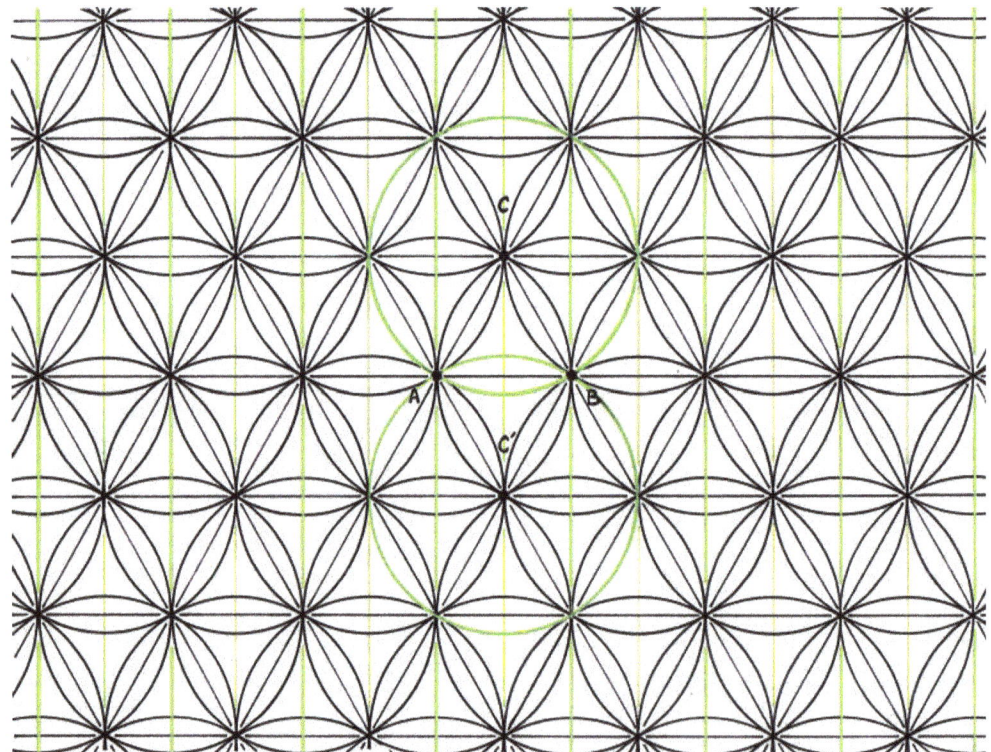

Figure 21b. Three √3 Spatial Frequency orientations, fundamental period.

Figure 4 also permits construction of a 'green' triangle-pair (and Figure 5 illustrates this) inscribed in the total space of the reference circle-pair. The green triangle-pair *shares* CC′ as a common side. The black triangle-pair and the green triangle-pair compose a trio of mutually perpendicular line-pair *complexes*.

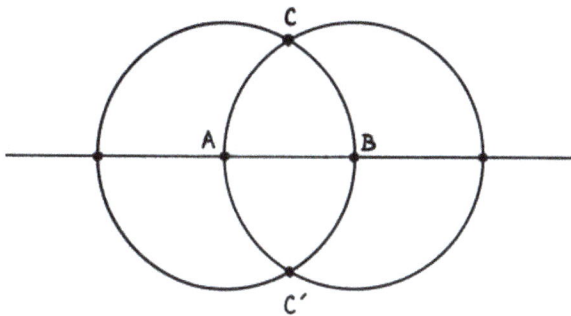

Figure 4.

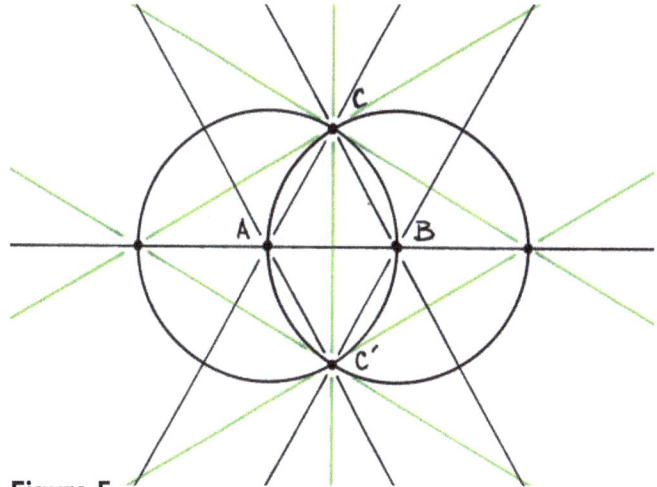

Figure 5.

In Figure 22, the black triangle-pair *tiles the plane*; the green triangle-pair *tiles the plane*. The superposition of this pair of infinite plane-tiles composes the *complex*: the trio of mutually ⊥ spatial frequency-pairs shown in Figure 22.

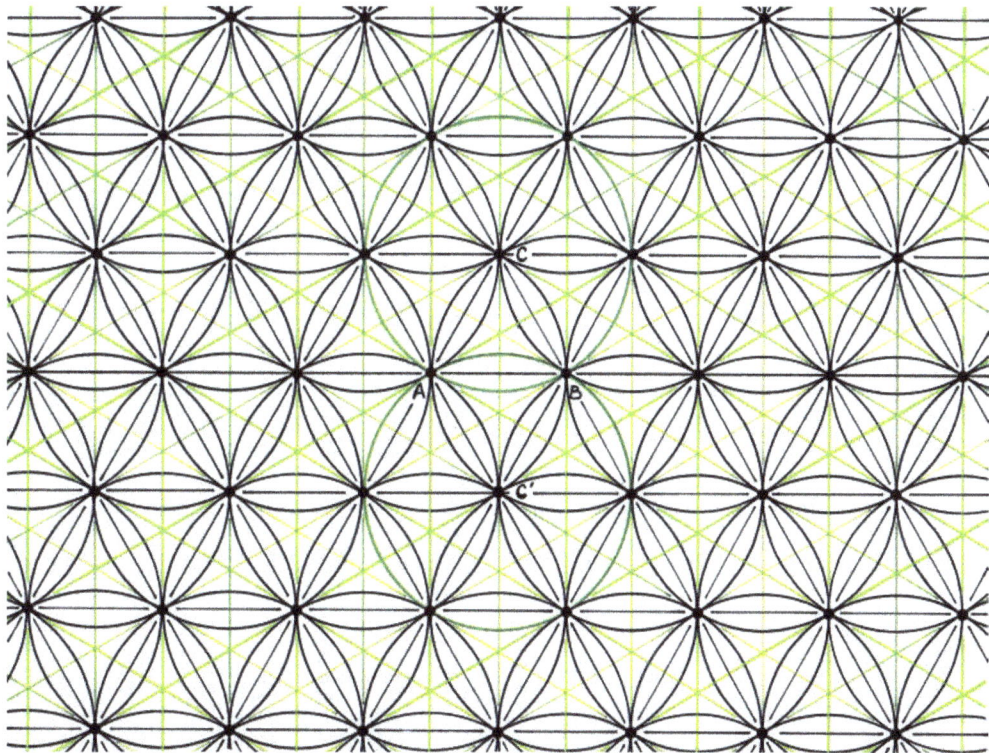

Figure 22. Three ⊥ pairs of √3 Spatial Frequency orientations, the fundamental period.

2. The superposed pattern of the mutually ⊥ *pair* of triangle-pair patterns permits a countable cyclic sequence of *divisions* of the areas of the black triangle-pair and the green triangle-pair.

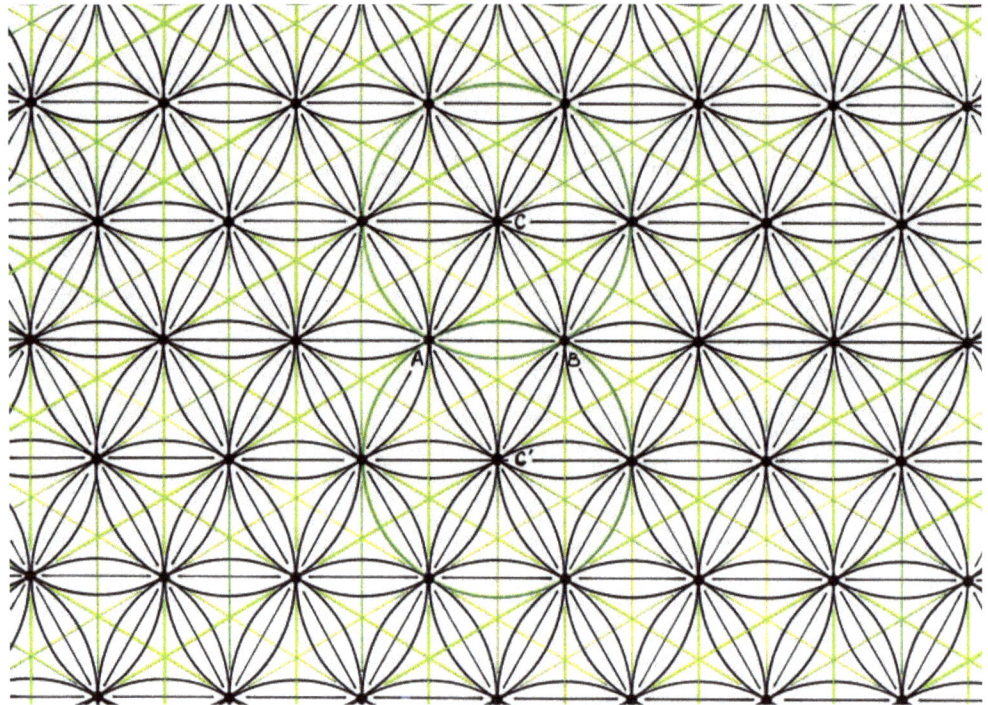

Figure 22. Three ⊥ pairs of √3 Spatial Frequency orientations, the fundamental period.

A finite number of these cycles of division is shown in Figure 23. This countable sequence of <u>*area*</u>-division pairs (one green, one black) <u>*constructs*</u> a planar pattern of *point locations* which specify, with <u>positional</u> <u>notation</u>, length measures of the countably infinite *powers* of the √3 basis number system, [16].

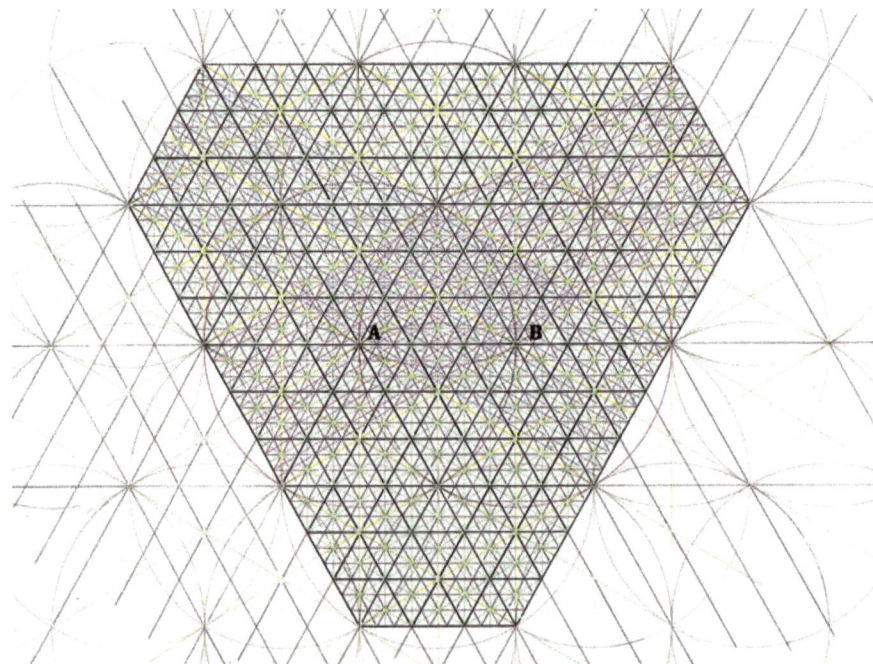

Figure 23. Original vintage drawing by Robert L. Powell, Sr. (See larger image p. 66.)

This development of the relationship between the articulated pair of triangle-pairs supports the Number Theory Theorem that: *'Three'* is the *first number*. (Figure 23)

3. The *complex*, the trio of black lines and the trio of green lines that pass through the reference point A (or B) *divides* the plane into twelve *equal* angle areas at the point A (or B). (Figure 23)

    Every circle *center* in Figure 19a is a canonical replication of the point A (or B). Hence, we have a planar *lattice* of fractal number system of *calibrations* of Euclid's plane.

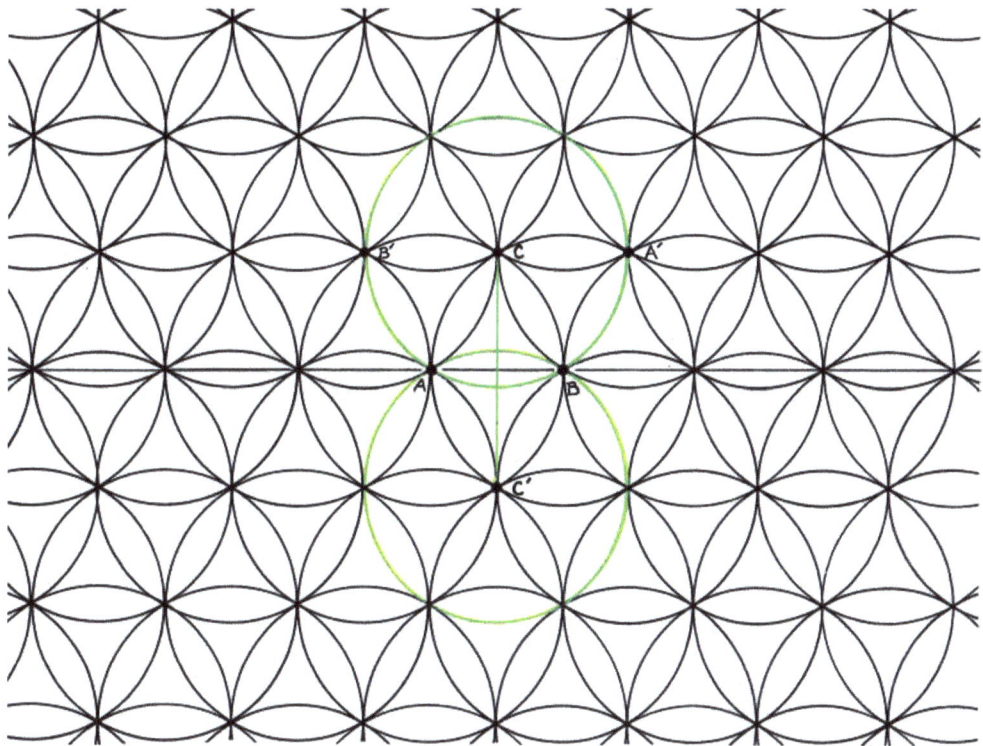

Figure 19a. A plane-filling replication of the shared radius.

B. The planar, √2 basis, fractal number system.

1. When the black trio of frequencies is introduced into the pattern of radius-sharing circles, Figure 19a, *no new points* are produced. When, on the other hand, the green trio of frequencies is introduced into the figure, *two patterns of new points* are produced:

(1) a pattern of points where [a certain] three green spatial frequency lines intersect; and

(2) a pattern of points where a green line segment intersects (perpendicularly) a *pair* of *circumferences*.

The two new point patterns, color-coded, are identified in Figure 24.

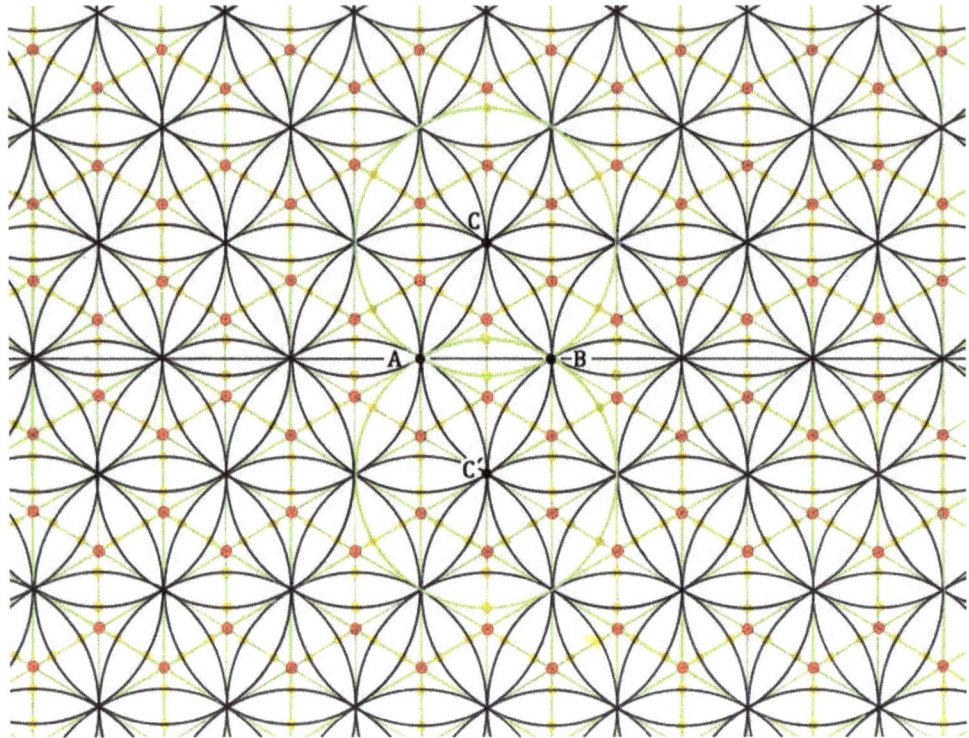

Figure 24. The [√3]⁻¹ point-pattern and the [√2] & [√2]⁻¹ lines-enabling point-pattern.

2. The pattern of 'red' points is the pattern of the '*three green line*' intersection points. The 'red' points, pair-wise, *trisect* the CC′ segment lengths. Since the CC′ length is determined as CC′ ≡ √3, the red point pattern computes the length,

$$(⅓) \, CC′ = (⅓)\sqrt{3} ≡ 1/\sqrt{3} = [\sqrt{3}]^{-1}.$$

This is the number required to structure the most primitive fractal number system basis, [16]. This number was constructed (computed) in Figure 11, as the radius of the '*Three*'-circle.

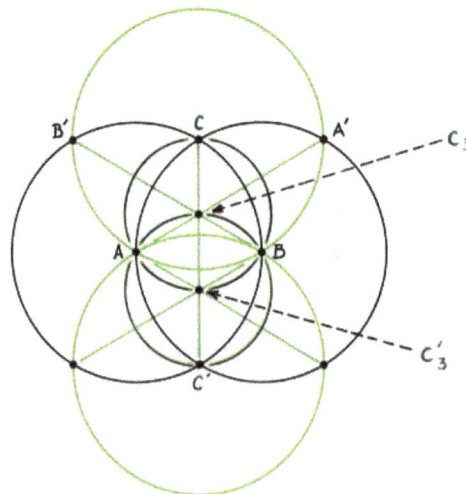

Figure 11.

3. The pattern of yellow-gold points in Figure 24 is formed by the _green line-circumference_ intersections.

    Our study of the use of this point-pattern by the guilds of ancient geometers _teaches_ that the pattern determines the system of a three-fold symmetric set of families of mutually ⊥ line-pairs. This three-family _complex_ of line-pairs is shown, color-coded, in Figure 25.

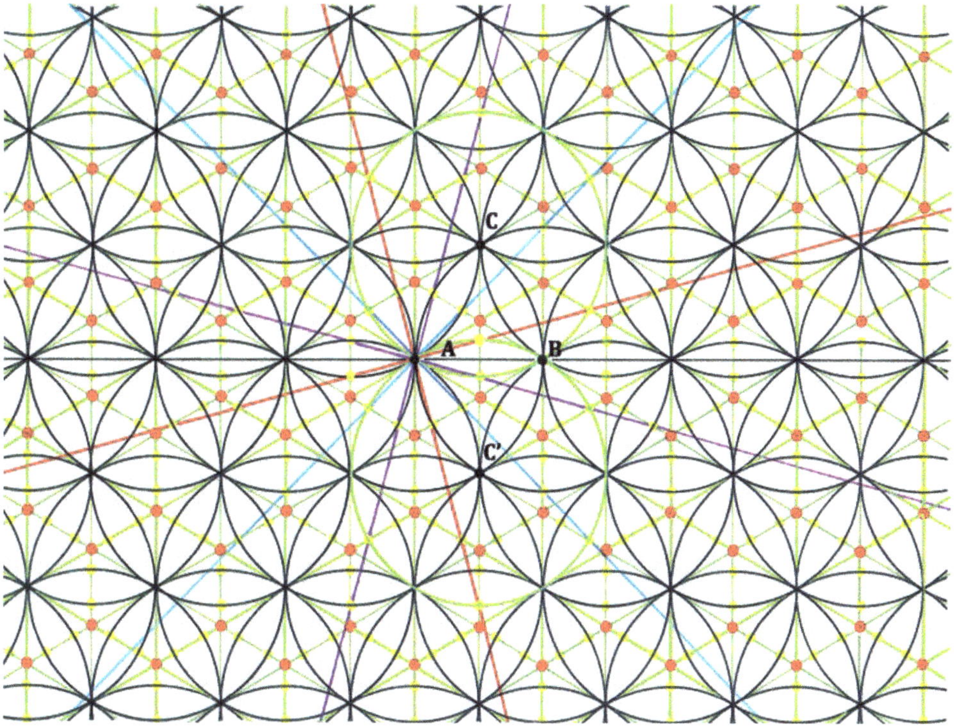

Figure 25. The principal √3 lines plus the principal √2 lines.

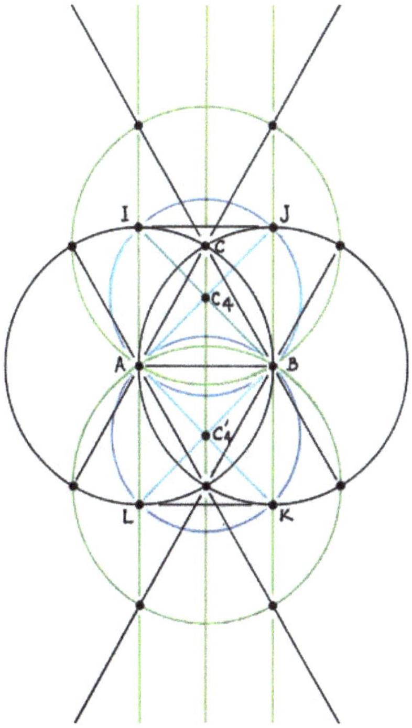

Figure 12.

This _complex_ of line-pairs is both necessary and sufficient for the construction/_computation_ of the fractal integer number-pair, $\sqrt{2}$ and $[\sqrt{2}]^{-1}$ {≡ ½√2}. (This number-pair was _computed_ in Figure 12, using the 'blue' color-coded line-pair, in the course of constructing/composing the 'Four'-circle.)

4. In Figure 12, the square-pair which shares the reference element AB, together with the pair of 'blue' diagonals, constitutes a <u>*complex*</u> which *tiles the plane*. The superposed pattern of (1) the square-pair replication pattern and (2) the pattern of replications of the pair of diagonal-pairs *permits* a countably infinite cyclic sequence of *divisions* of the <u>areas</u> of the square-pairs. A finite number of cycles of this sequence of divisions is shown in Figure 26.

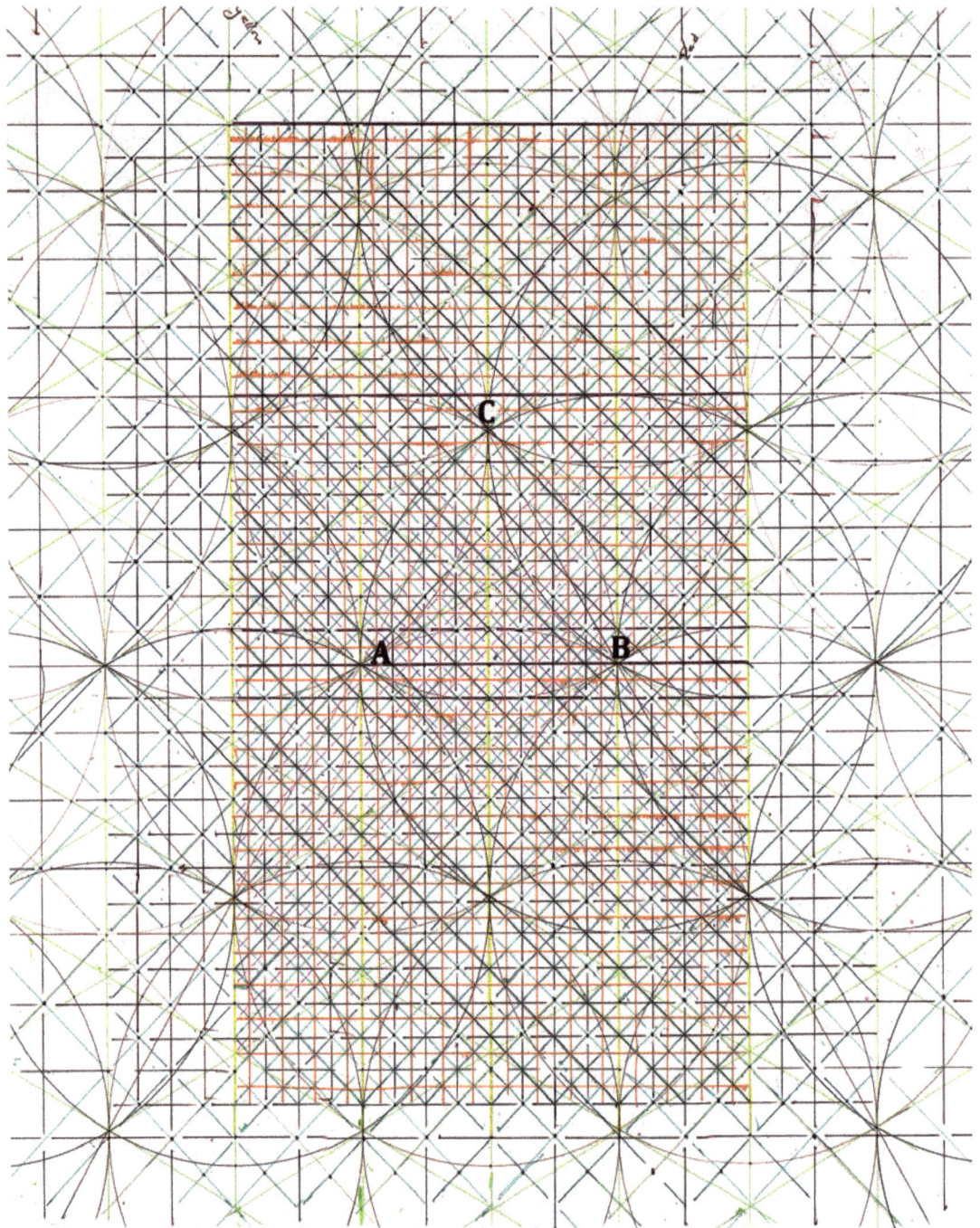

Figure 26. The *planar* space-filling √2 number fractal co-ordinate system. Original vintage drawing by Robert L. Powell, Sr.

This sequence of divisions of the fundamental (reference) square-pair <u>area</u> constructs a planar pattern of *point locations* which compute, in <u>positional</u> <u>notation</u>, the sequence of countably infinite *powers* of $[\sqrt{2}]^1$: the second-most primitive fractal integer base number system [17]. Thus, '*two*' is the *second* number.

5. The color-coded *complex* of '$\sqrt{2}$' line-pairs in Figure 25 also <u>*bisects*</u> the twelve equal angles made by the *complex* of mutually $\perp$ black line-green line '$\sqrt{3}$' lines in Figure 22. Together, they divide the plane into twenty-four equal angle areas, at the center of every circle in the lattice replication.

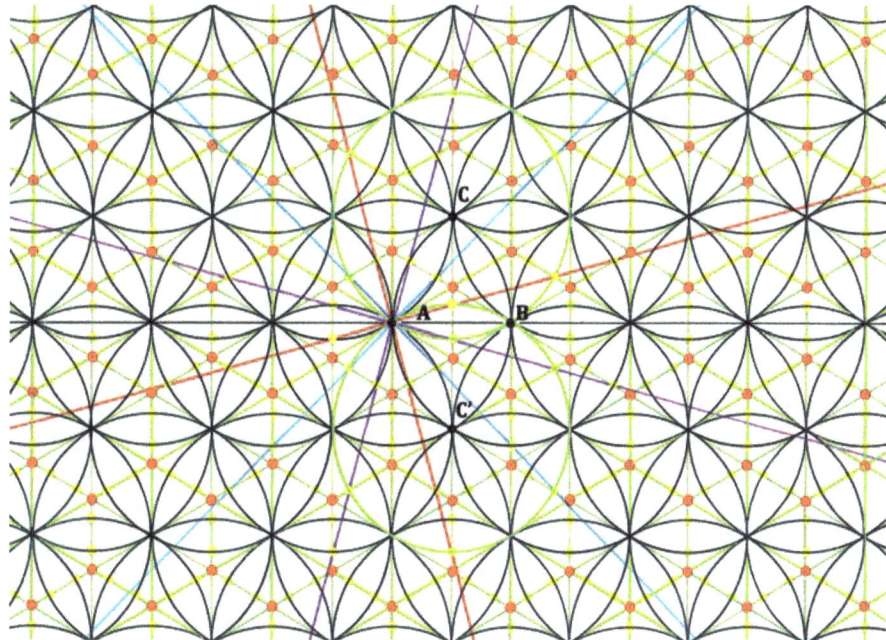

Figure 25. The principal √3 lines plus the principal √2 lines.

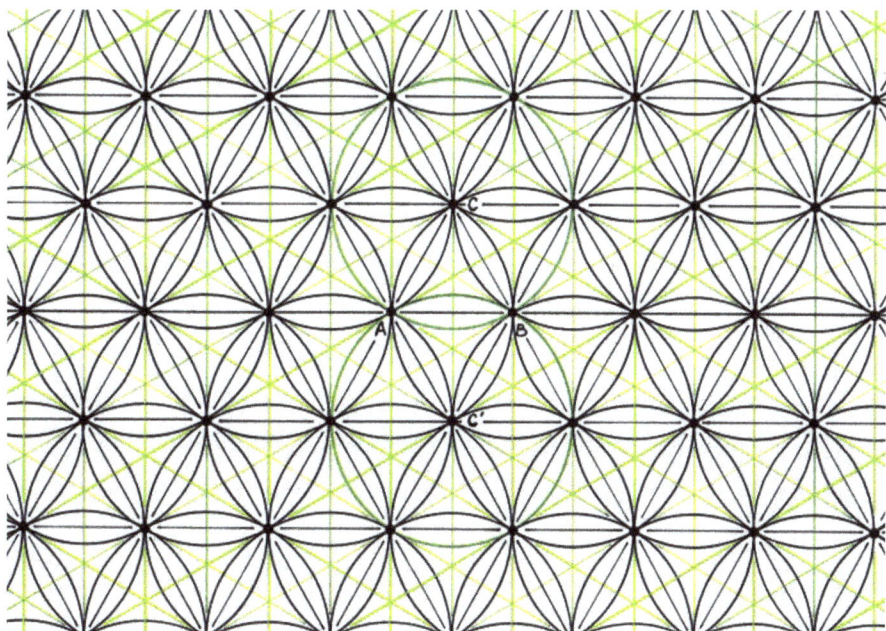

Figure 22. Three ⊥ pairs of √3 Spatial Frequency orientations, the fundamental period.

C. A *third* **class of new points** is created, when the black trio of replications of the reference length AB and the mutually ⊥, mutually *bisecting* CC′ replications are replicated, shown in Figure 22.

D. <u>The planar, ½ [√5 ± √1] basis, fractal number system.</u>

1. The pattern of yellow-gold points depicted in Figure 24 is the *architecture* that is both necessary and sufficient for a canonical construction of the ***space-filling*** Fractal Number positional notation co-ordinate system calibrated in countably infinite *powers* of ½[√5 ± √1], i.e., the basis number system, [18].

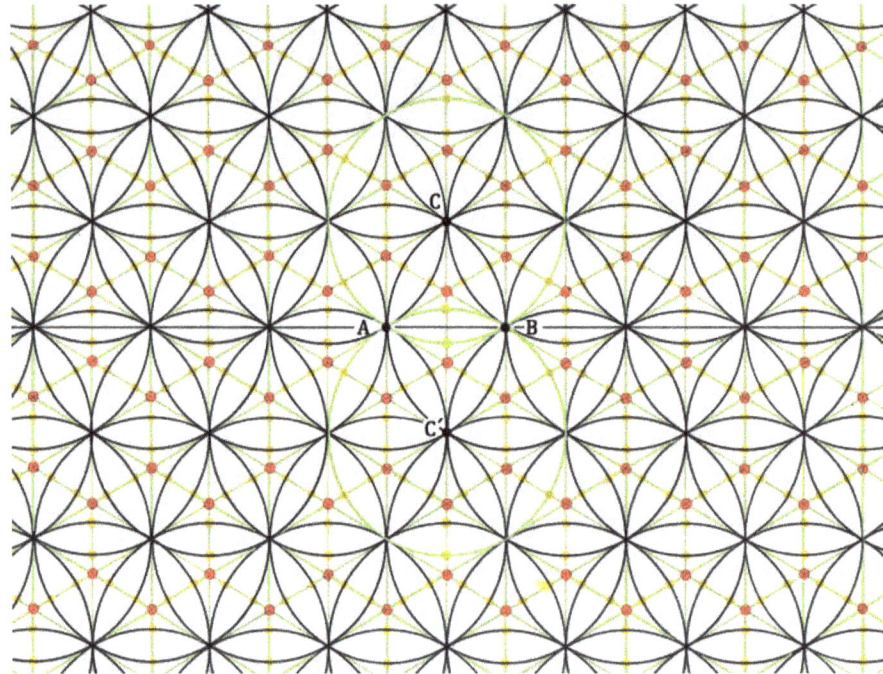

Figure 24. The [√3]⁻¹ point-pattern and the [√2] & [√2]⁻¹ lines-enabling point-pattern

This *architecture* was employed in the Figure 13 construction of the strategic 'Orange' circle. It was determined so as to *circumscribe* the reference square-pair, [I, J, K, L]. This computed the (directed) length segments OM & ON, co-linear with the length segments OA & OB.

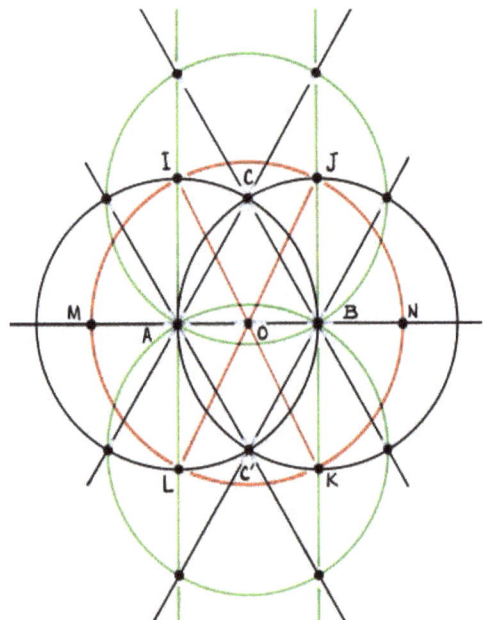

Figure 13. The "Orange' circle, radii ½ [IK]

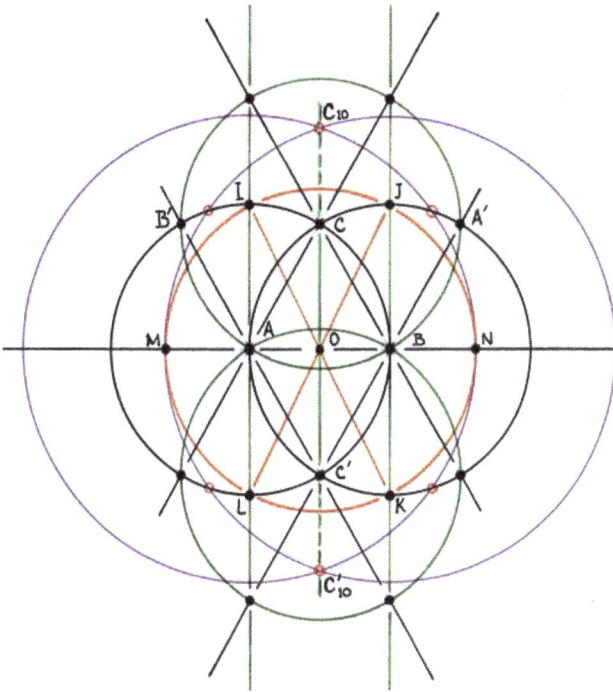

Figure 14. Purple circle-pair, radii = Φ [=½(...+...)]

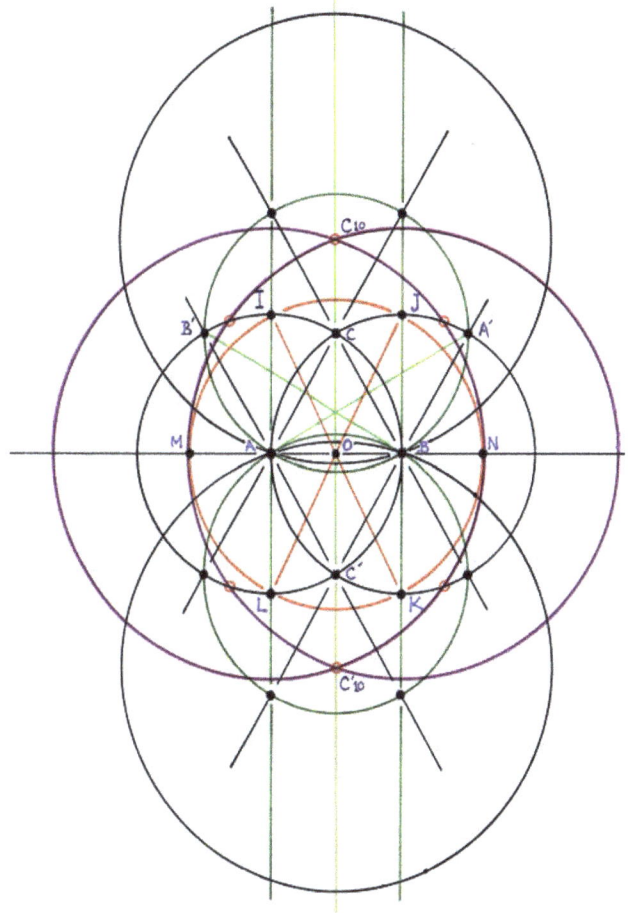

Figure 15. Black circle-pair, radii = Φ

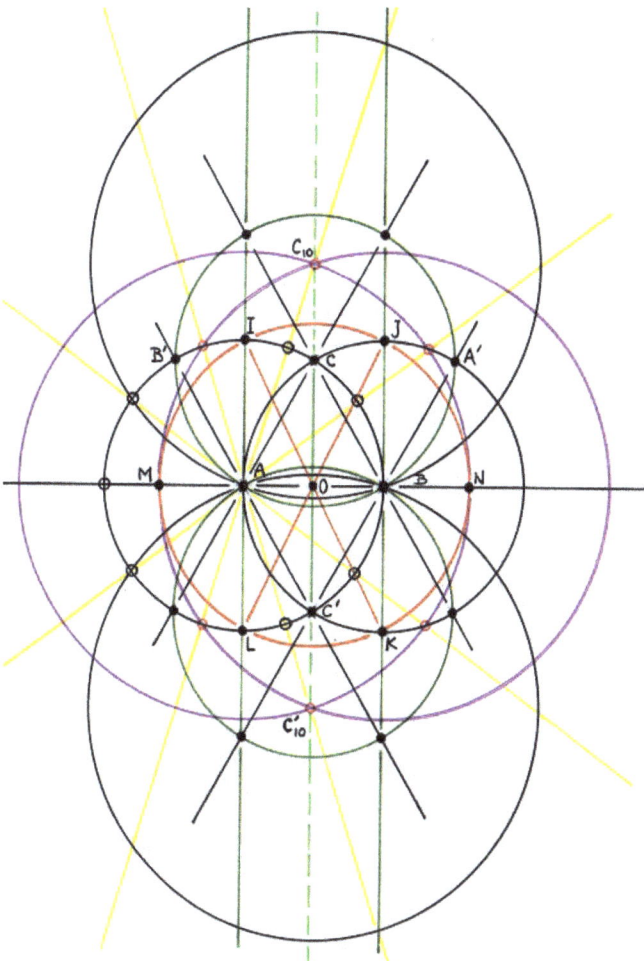

Figure 27. The principal ½[√5±√1] lines.

Co-linear combinations of this quartet of segments compute the radii of the *purple* circle-pair, in the Figures 14 & 15: radii = OA + ON = OB + OM ≡ ½[IK + AB] ≡ ½[√5 + √1].

The big black circle-pair in Figure 15 shares a radius with the purple circle-pair: radius of the big black circle-pair in Figure 15 - ½ [√5 + √1], also.

2. Figure 27 identifies a pair of trios of points in the complex, see Figure 15, color-coded 'red'.

Two remarks about this pair of trios:

(a) together with the A point, they determine the quartet of yellow-gold lines shown passing through the A point [they also determine a similar quartet of line-family *complex* passing through the B point]; and

(b) together with the reference binary point-pair, they determine a regular pentagon-***pair***, with AB as a *shared* side.

Figure 27 shows this quartet of determined yellow-gold lines.

Together with the AB line extensions, this complex of five 'pentagon-***pair*** lines' divides the plane at the point A, into *ten* equal angle zones.

3. Figure 27 shows the pentagon-pair, with shared side AB, determined by the pair of trios of 'red' points in Figure 27, together with the binary point-pair reference.

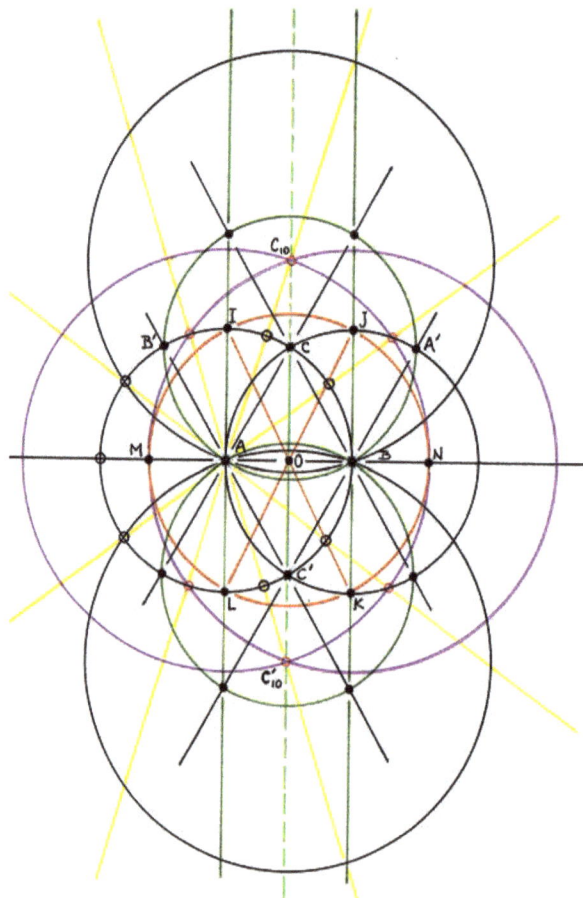

Figure 27. The principal ½[√5±√1] lines.

a. This pentagon-pair embeds an isosceles triangle-pair, sides = ½[√5 - √1] = {½ [√5 + √1]}⁻¹, with shared side = AB = √1.

b. This pentagon-pair embeds another isosceles triangle-pair, sides = ½[√5 + √1], with shared side = AB = √1.

c. This pentagon-pair permits construction of the hexagonal regular polytope[12], shown in Figures 28a and 28b, with width = {½[√5 + √1]}³.

4.   Each isosceles triangle-pair *tiles the plane*.

5.   The hexagonal regular polytope, Figures 28a and 28b, ***tiles** the Euclidean plane*.

6.   The five pentagon lines through the A point (and/or through the B point) constitute the *mathematical complex* which calibrates the Euclidean plane with *areal* positional notation, in *powers* of the fractal number basis system [18].

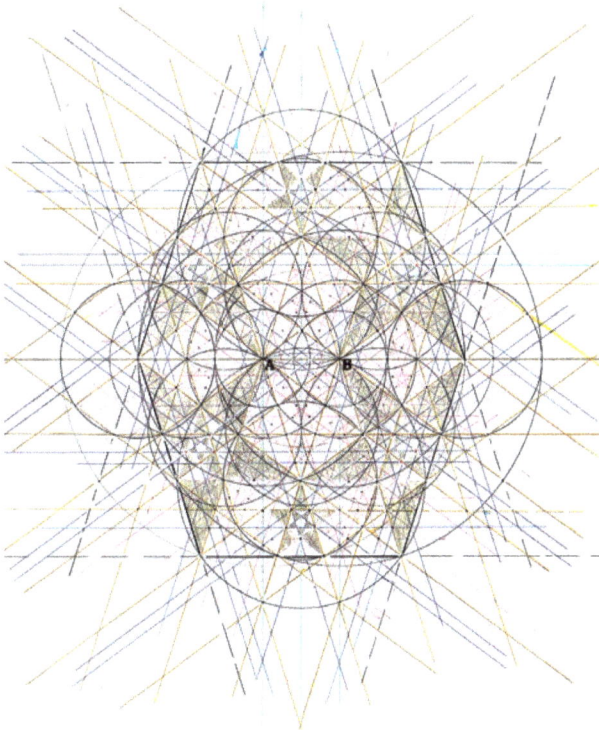

Figure 28a. Original vintage drawing by Robert L. Powell, Sr. (See larger image on p. 16.)

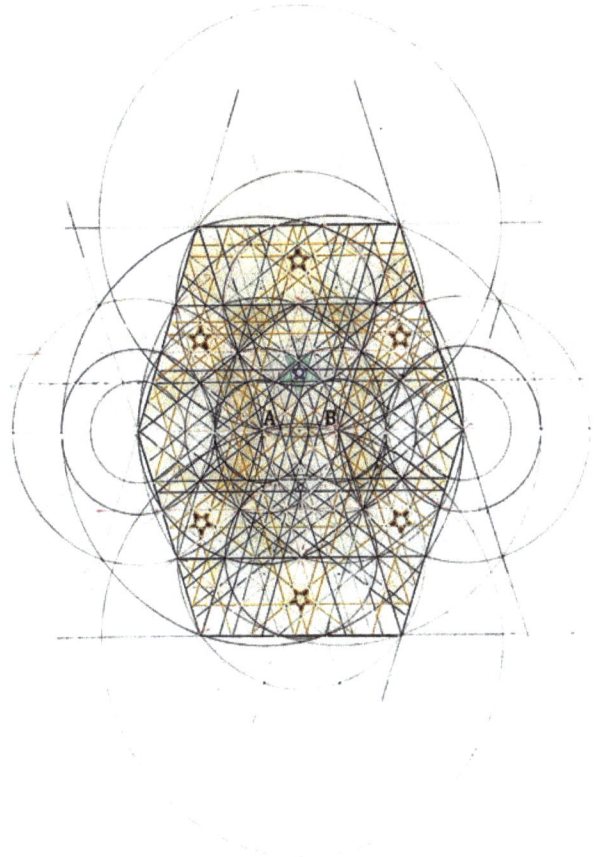

Figure 28b. The planar space-filling ½[√5±√1] fractal number co-ordinate system. Original vintage drawing by Robert L. Powell, Sr. (See larger image on p. 17.)

9   Goldman, Jay R. *The Queen of Mathematics: A Historically Motivated Guided to Number Theory*. A. K. Peters Ltd. Wellesley Mass (1998), p. 204.
10 Ibid.
11 Newman, James R. *The World of Mathematics*. Simon and Schuster (1956), vol. 1, p. 237
12 H. S. M. Coxeter, *Regular Polytopes*. Dover Publications, New York (1973).

Chapter VI   THE HINNANT-POWELL FRACTAL NUMBER SERIES:
A *Visual Mathematics* Discovery

A. In the foregoing chapters, we have developed a parallel-organized system of new and *evolutionary* generalizations of the Epistemology of 20th century mathematics, based on the *first* of a countably infinite number of entry steps directly into a mathematizable domain of transcendental number and its corollary domain of trans-rational algebraic number arithmetics.

   The generalizations emerged as the logical consequences of a *revolutionary* definition of the measure of the Euclidean interval, AB, determined by an arbitrary binary point-pair, [A, B]:

$$AB \equiv [1]^{1/n}, \qquad n = 1, 2, 3, \ldots, \qquad [1]'$$

together with a *complete* employment of the implications of the *invariant* mutual corollary with AB, namely CC′.

B.1. Chapter I permitted a periodic lattice of polar coordinate calibrations in terms of the transcendental number integers, [6] and [7], and in terms of the trans-rational number integers, [8]. This result derived from the inter-articulation of the 'green'-line family with the 'black'-line family.

B.2. Chapter II, similarly, permitted the periodic lattice of polar coordinate calibrations in terms of powers of the three Platonic fractal number integers, fundamental to Gauss' Cyclotomy problem. These results were permitted when the '$\sqrt{3}$' line-pair *complex* is employed as a *necessary* platform on which to construct the '$\sqrt{2}$' *complex* of line-pairs.

B.3. Chapter IV produced a two-dimensional, infinite table of fractal number integer Trigonometry Ratios—as the *complete* set of solutions to a canonical system of transcendental number integer equations. This set of transcendental number equations was constructed on the structural foundation of the AB line and the family of green lines perpendicular to it.

C. This chapter employs the superposition of the *complete* set of green-line/black-line families of perpendicular pairs and the *complete* set of '$\sqrt{2}$' 'diagonals' perpendicular line-pairs. The parallel-organized superposition of this pair of *complexes* of trios of perpendicular line-pair *complexes* is shown in Figure 25.

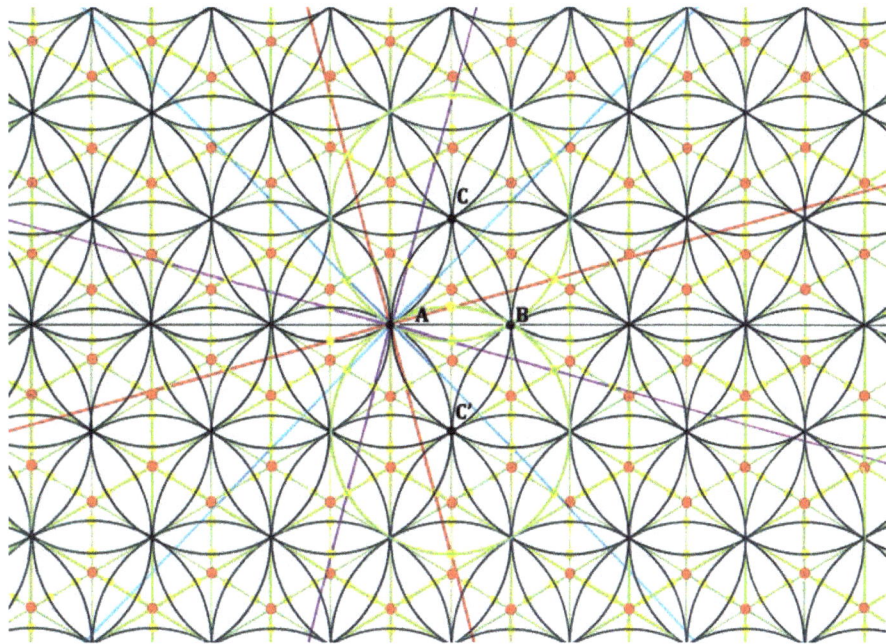

Figure 25. The principal √3 lines plus the principal √2 lines.

1. This family of lines is composed of six pairs of mutually ⊥ line-pairs: a blue ⊥ line pair, a red ⊥ line pair, a purple ⊥ line pair; and three mutually ⊥ green-line/black-line pairs.

    The green-line/black-line pairs 'inform' the A point of the entire plane of potential point-locations, calibrated in the powers of $[\sqrt{3}]^{\pm1}$, in the three-fold symmetry of CC', AA', BB'.

    The blue-line, red-line, and purple-line pairs inform the A point of the entire plane of potential point-locations in terms of positional notation in the base number, $[\sqrt{2}]$. The vector orientation of the √2 calibrations derives from the arbitrary AB-CC' reference orientation.

2. The laws of Figure 3 permit the selection of the B point in Figure 2 to *immediately* determine a countably infinite plane-filling (face-centered hexagonal) array of circle *centers*, replications of the binary point-pair reference.

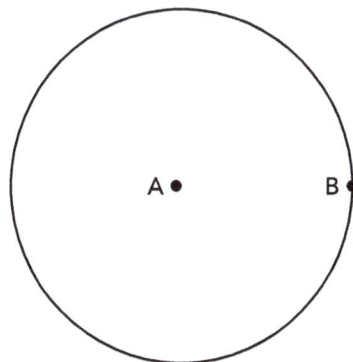

Figure 3a.

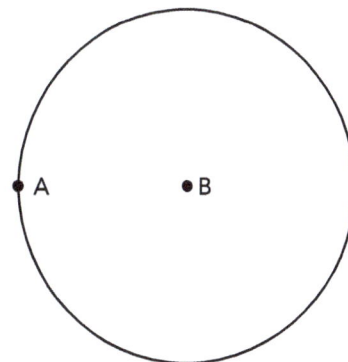

Figure 3b.

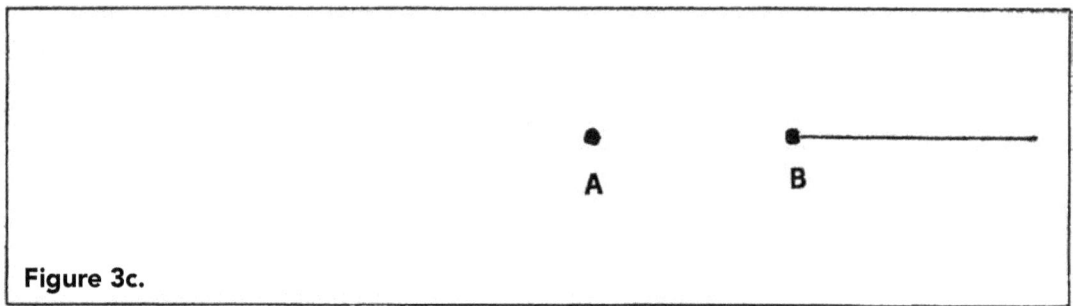

Figure 3c.

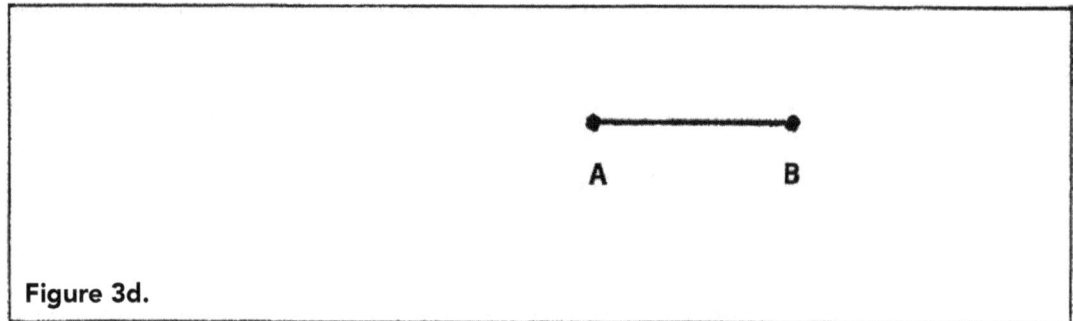

Figure 3d.

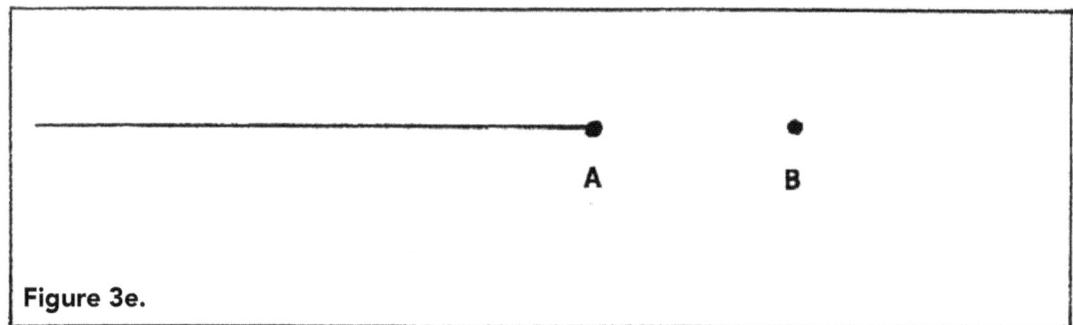

Figure 3e.

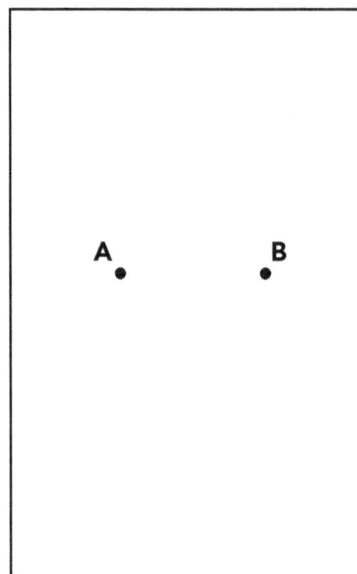

Figure 2. A reference binary pair, [A,B], giving coordinate calibration to the *entire* Euclidean *number plane*.

Each circle center in the fractal lattice of centers is informed of the entire plane of potential point coordinates by a *quantized* space-shift replication of the six pairs of mutually ⊥ line pairs converging at the A point in Figure 25.

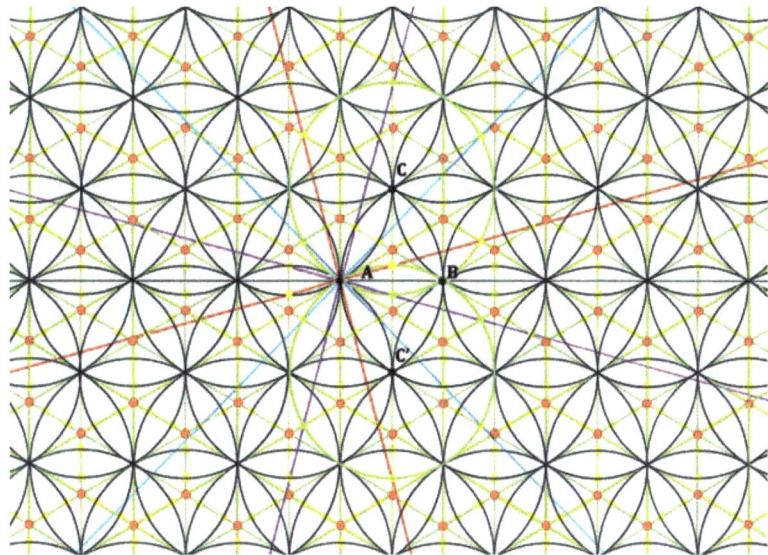

Figure 25. The principal √3 lines plus the principal √2 lines.

3. *After* the fact of a serendipitous discovery, it should not surprise one that this ubiquitous *complex* of lines should yield an efficient and elegant package of significant *fractal number computations*—packages involving lovely combinatorials of [±] powers of the *symmetry architectures* of √1, √3, and √2—the fractal number quantizations of the plane space-filling triangle-pair; its derivative, the plane space-filling square-pair; and the square-pair's derivative, the plane space-filling regular polytopes with space quantum numbers constrained to powers of ½[√5 ± √1].

   This chapter introduces an after-the-fact description of just such a serendipitous discovery.

4. To begin inquiry, we enrich the background of Figure 29a for a quadrant of the 24 rays from the A point.

Figure 29a. Platform for Hinnant-Powell fractal number systems.

Figure 29b shows a first & third quadrant representation of the complete set of √3 perpendicular line-pairs and the complete set of √2 'diagonals' perpendicular line-pairs.

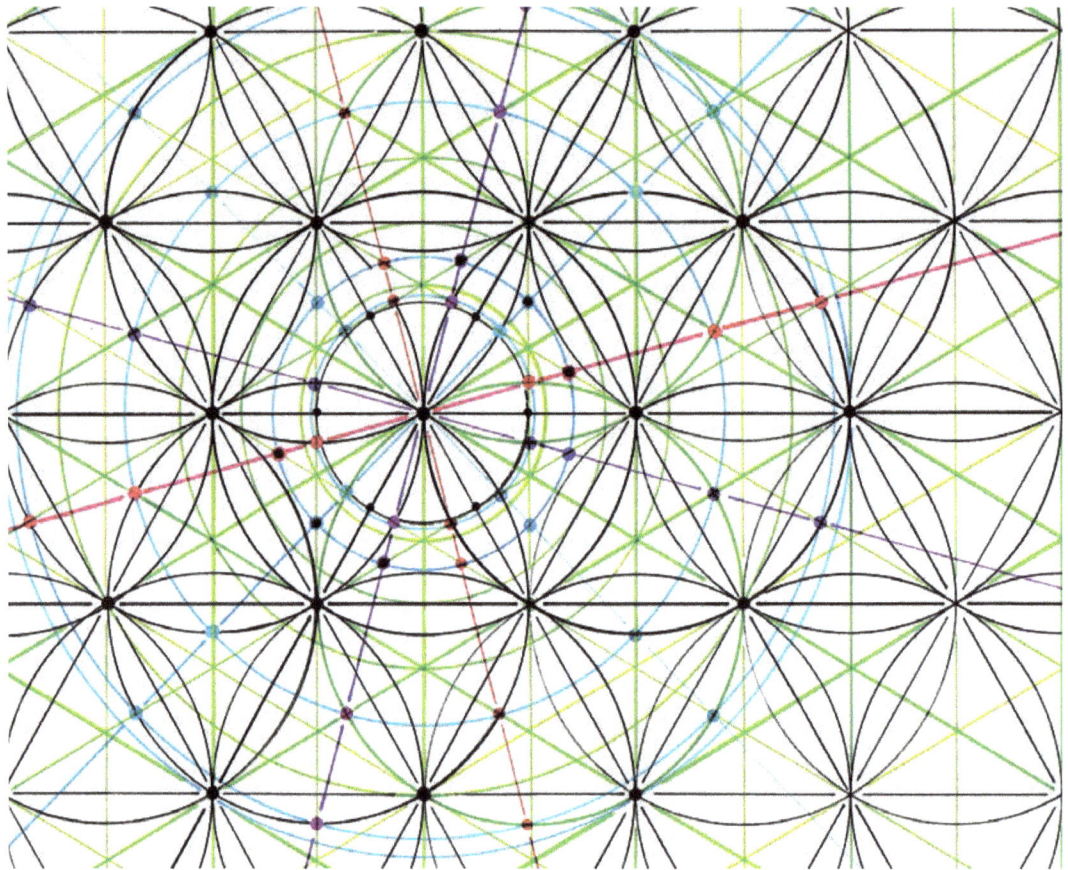

Figure 29b. Platform for Hinnant-Powell number systems, 1st and 3rd quadrants.

On this structure foundation, we introduce a system of known *circumferences* concentric at A, with *radii* =

½√1 (black),
√(√4 - √3) (blue, dashed line),
(⅓)√3 (green, dashed line),
½√2 (blue).
½√3 (green),
√1 (black),
(⅔)√3 (green),
√2 (blue),
√3 (green),
√2 + √(√4 - √3) (blue),
√4 (blue).

This *complex* of circumferences is shown, also, in Figure 29c.

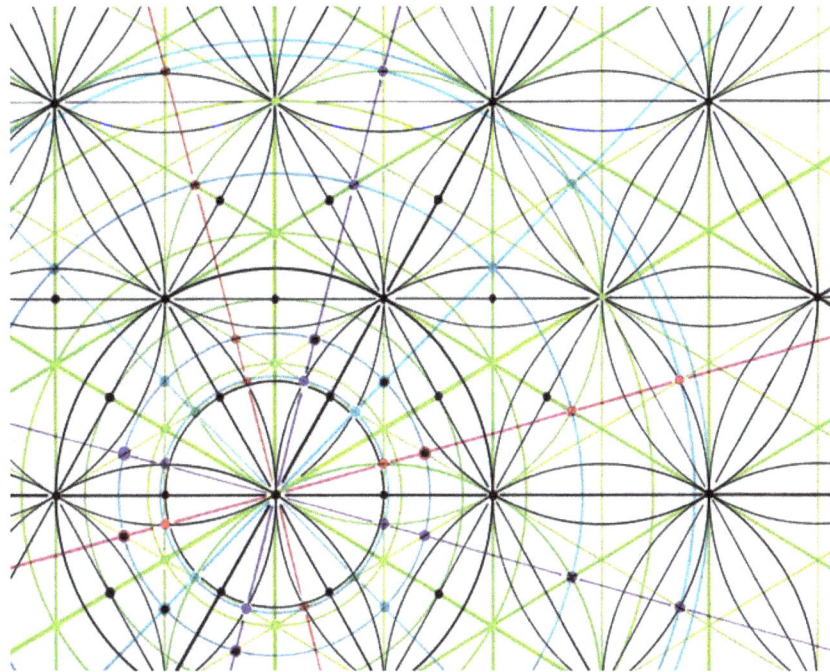

Figure 29c. Platform for Hinnant-Powell fractal number systems, 1st quadrant emphasis.

5. This system of *circumferences* marks its system of *radii* on each of the (seven) (color-coded) rays through the A point of Figure 29c in the quadrant.

6. Poring over the interaction between the seven rays and the eleven circumferences, one of us (V.H.) grokked the proposition that <u>some</u> of the ray lines were *sectioned*, by the background of *circumferences*, into a 'da Vinci code' ratio of *segments*.

   Hinnant's speculative *gestalt sensibility* grokked—indeed demanded—that we examine the <u>computed</u> validity of the segment-ratios highlighted in Figure 29d.

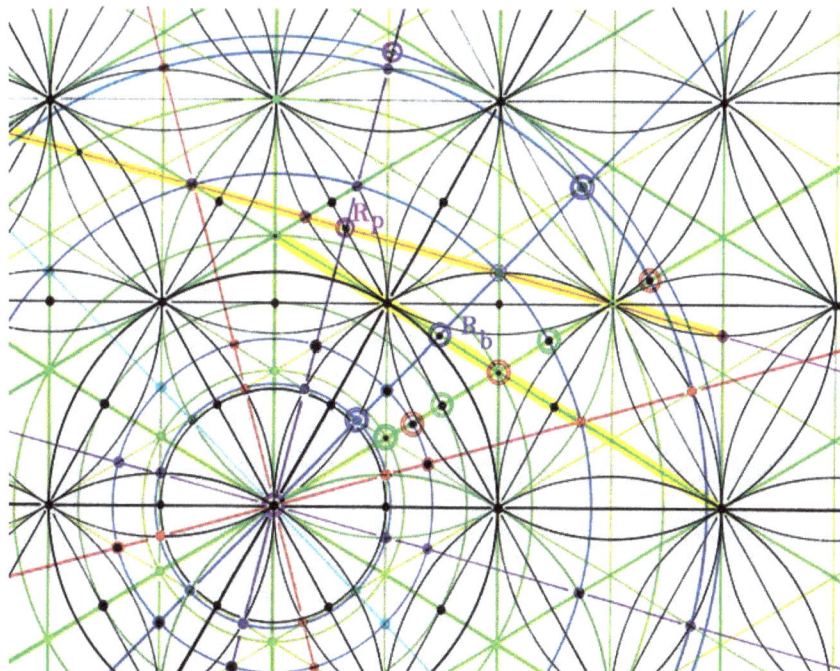

Figure 29d. The *visually identified* Da Vinci ratio *sections*.

The following is an EXPLORATORY investigation to assess if there is a precise or approximate relationship of the Golden proportion in these candidates.

Expressed in terms of the eleven circumference radii shown in Figure 29d, the highlighted candidates for 'da Vinci code' ratio segments are:

radial line, color-coded:	larger section	smaller section	total length
1. purple	R_{purple}	$\sqrt{4} - R_{purple}$	$\sqrt{4}$
2. blue	$\sqrt{2} + \sqrt{(\sqrt{4} - \sqrt{3})} - R_{blue}$	$R_{blue} - \sqrt{(\sqrt{4} - \sqrt{3})}$	$\sqrt{2}$
3. green	$\sqrt{2} - \frac{1}{2}\sqrt{3}$	$\frac{1}{2}\sqrt{3} - (\frac{1}{3})\sqrt{3}$	$\sqrt{2} - (\frac{1}{3})\sqrt{3}$
4. red	$\sqrt{2} + \sqrt{(\sqrt{4} - \sqrt{3})} - (\frac{2}{3})\sqrt{3}$	$(\frac{2}{3})\sqrt{3} - \frac{1}{2}\sqrt{2}$	$\frac{1}{2}\sqrt{2} + \sqrt{(\sqrt{4} - \sqrt{3})}$

Table Ia.

It can be shown that the *geometric* <u>computation</u> for R_{purple} and R_{blue} are, by *visual* inspection,

$$R_{purple} = \frac{1}{2}(\sqrt{3})(\sqrt{2})[= \frac{1}{2}\sqrt{6}], \text{ and}$$

$$R_{blue} = 2[\sqrt{(\sqrt{4} - \sqrt{3})}).$$

Substituting these values gives:

radial line, color-coded:	larger section	smaller section	total length
1. purple	$\frac{1}{2}\sqrt{6}$	$\sqrt{4} - \frac{1}{2}\sqrt{6}$	$\sqrt{4}$
2. blue	$\sqrt{2} - \sqrt{(\sqrt{4} - \sqrt{3})}$	$\sqrt{(\sqrt{4} - \sqrt{3})}$	$\sqrt{2}$
3. green	$\sqrt{2} - \frac{1}{2}\sqrt{3}$	$\frac{1}{2}\sqrt{3} - (\frac{1}{3})\sqrt{3}$	$\sqrt{2} - (\frac{1}{3})\sqrt{3}$
4. red	$\sqrt{2} + \sqrt{(\sqrt{4} - \sqrt{3})} - (\frac{2}{3})\sqrt{3}$	$(\frac{2}{3})\sqrt{3} - \frac{1}{2}\sqrt{2}$	$\frac{1}{2}\sqrt{2} + \sqrt{(\sqrt{4} - \sqrt{3})}$

Table Ib.

7. Each of this quartet of sectioned total length segments is the generator of (a) an *ever*-increasing and (b) [oscillatory] ever-*decreasing* Fractal Number generalization of the 'Golden Section' self-replication series' property. The computations of these generalizations are collected in TABLE II and TABLE III.

Each quartet is the generator of a fractal number generalization of the Fibonacci scale of self-replications.

purple:		blue:	
ever-increasing:	**ever-decreasing:**	**ever-increasing:**	**ever-decreasing:**
{S:} $\sqrt{4}-\frac{1}{2}(\sqrt{2})(\sqrt{3})$	{L:} $\frac{1}{2}\sqrt{6}$	{S:} $\sqrt{(\sqrt{4}-\sqrt{3})}$	{L:} $\sqrt{2}-\sqrt{(\sqrt{4}-\sqrt{3})}$
{L:} $\frac{1}{2}\sqrt{6}$	{S:} $\sqrt{4}-\frac{1}{2}\sqrt{6}$	{L:} $\sqrt{2}-\sqrt{(\sqrt{4}-\sqrt{3})}$	{S:} $\sqrt{(\sqrt{4}-\sqrt{3})}$
——————:	——————:	——————:	——————:
{T:} $\sqrt{4}$	$-\sqrt{4}+2\sqrt{6})/2$	{T:} $\sqrt{2}$	$\sqrt{2}-2\sqrt{(\sqrt{4}-\sqrt{3})}$
$\sqrt{4}+\frac{1}{2}\sqrt{6}$	$2\sqrt{4}-3\frac{1}{2}(\sqrt{2})(\sqrt{3})$	$2\sqrt{2}-\sqrt{(\sqrt{4}-\sqrt{3})}$	$\sqrt{2}+3\sqrt{(\sqrt{4}-\sqrt{3})}$
$2\sqrt{4}+\frac{1}{2}\sqrt{6}$	$-3\sqrt{4}+5(\sqrt{6})/2$	$3\sqrt{2}-\sqrt{(\sqrt{4}-\sqrt{3})}$	$2\sqrt{2}-5\sqrt{(\sqrt{4}-\sqrt{3})}$
$3\sqrt{4}+2(\frac{1}{2}\sqrt{6})$	$+[5\sqrt{4}-(\frac{1}{2}\sqrt{6})]$	$5\sqrt{2}-2\sqrt{(\sqrt{4}-\sqrt{3})}$	$-3\sqrt{2}+8\sqrt{(\sqrt{4}-\sqrt{3})}$
$5\sqrt{4}+3(\frac{1}{2}\sqrt{6})$	$-[8\sqrt{4}-(\frac{1}{2}\sqrt{6})]$	$8\sqrt{2}-3\sqrt{(\sqrt{4}-\sqrt{3})}$	$5\sqrt{2}-13\sqrt{(\sqrt{4}-\sqrt{3})}$
$8\sqrt{4}+5(\frac{1}{2}\sqrt{6})$	$+[13\sqrt{4}-(\frac{1}{2}\sqrt{6})]$	$13\sqrt{2}-5\sqrt{(\sqrt{4}-\sqrt{3})}$	$-[8\sqrt{2}-21\sqrt{(\sqrt{4}-\sqrt{3})}]$
$13\sqrt{4}+8(\frac{1}{2}\sqrt{6})$	$-[21\sqrt{4}-(\frac{1}{2}\sqrt{6})]$	$21\sqrt{2}-8\sqrt{(\sqrt{4}-\sqrt{3})}$	$+[13\sqrt{2}-34\sqrt{(\sqrt{4}-\sqrt{3})}]$
$21\sqrt{4}+13(\frac{1}{2}\sqrt{6})$	$+[34\sqrt{4}-(\frac{1}{2}\sqrt{6})]$	$34\sqrt{2}-13\sqrt{(\sqrt{4}-\sqrt{3})}$	$-[21\sqrt{2}-55\sqrt{(\sqrt{4}-\sqrt{3})}]$
$34\sqrt{4}+21(\frac{1}{2}\sqrt{6})$	$-[13\sqrt{4}-(34/2)\sqrt{6}]$	$55\sqrt{2}-21\sqrt{(\sqrt{4}-\sqrt{3})}$	$+[13\sqrt{2}-34\sqrt{(\sqrt{4}-\sqrt{3})}]$
…	…	…	…
…	…	…	…
…	…	…	…

Table II. The Ever-increasing & Ever-decreasing _pairs_ of The Fractal Integer Number Generalizations of The Fibonacci Number series of Sectionings

green:
ever-increasing:

{S:} $[\tfrac{1}{2}-(\tfrac{1}{3})]\sqrt{3}$

{L:} $\sqrt{2}-\tfrac{1}{2}\sqrt{3}$

——————:

{T:} $\sqrt{2}-[(\tfrac{1}{3})]\sqrt{3}$

$\quad 2\sqrt{2}-[\tfrac{1}{2}+(\tfrac{1}{3})]\sqrt{3}$

$\quad 3\sqrt{2}-[\tfrac{1}{2}+(\tfrac{2}{3})]\sqrt{3}$

$\quad 5\sqrt{2}-[(\tfrac{2}{2})+(\tfrac{3}{3})]\sqrt{3}$

$\quad 8\sqrt{2}-[(\tfrac{3}{2})+(\tfrac{5}{3})]\sqrt{3}$

$\quad 13\sqrt{2}-[(\tfrac{5}{2})+(\tfrac{8}{2})]\sqrt{3}$

$\quad$ …

red:
ever-increasing:

{S:} $-\tfrac{1}{2}\sqrt{2}+(\tfrac{2}{3})\sqrt{3}$

{I:} $\sqrt{2}-(\tfrac{2}{3})\sqrt{3}+\sqrt{(\sqrt{4}-\sqrt{3})}$

——————————:

{T:} $\tfrac{1}{2}\sqrt{2}+\sqrt{(\sqrt{4}-\sqrt{3})}$

$(\tfrac{3}{2})\sqrt{2}-(\tfrac{2}{3})\sqrt{3}+2\sqrt{(\sqrt{4}-\sqrt{3})}$

$(\tfrac{4}{2})\sqrt{2}-(\tfrac{2}{3})\sqrt{3}+3\sqrt{(\sqrt{4}-\sqrt{3})}$

$(\tfrac{7}{2})\sqrt{2}-2(\tfrac{2}{3})\sqrt{3}+5\sqrt{(\sqrt{4}-\sqrt{3})}$

$(\tfrac{11}{2})\sqrt{2}-3(\tfrac{2}{3})\sqrt{3}+8\sqrt{(\sqrt{4}-\sqrt{3})}$

$(\tfrac{18}{2})\sqrt{2}-5(\tfrac{2}{3})\sqrt{3}+13\sqrt{(\sqrt{4}-\sqrt{3})}$

green:
ever-decreasing:

{L:} $\sqrt{2}-\tfrac{1}{2}\sqrt{3}$

{S:} $[\tfrac{1}{2}-(\tfrac{1}{3})]\sqrt{3}$

——————:

$\sqrt{2}-[(\tfrac{2}{2})+(\tfrac{1}{3})]\sqrt{3}$

$-\sqrt{2}+[(\tfrac{3}{2})-(\tfrac{2}{3})]\sqrt{3}$

$2\sqrt{2}-[(\tfrac{5}{2})-(\tfrac{3}{3})]\sqrt{3}$

$-3\sqrt{2}+[(\tfrac{8}{2})-(\tfrac{5}{3})]\sqrt{3}$

$5\sqrt{2}-[(\tfrac{13}{2})-(\tfrac{8}{3})]\sqrt{3}$

$-8\sqrt{2}+[(\tfrac{21}{2})-(\tfrac{13}{3})]\sqrt{3}$

…

red:
ever-decreasing:

{L:} $\sqrt{2}-(\tfrac{2}{3})\sqrt{3}+\sqrt{(\sqrt{4}-\sqrt{3})}$

{S:} $-\tfrac{1}{2}\sqrt{2}+(\tfrac{2}{3})$

——————————:

$(\tfrac{3}{2})\sqrt{2}-2(\tfrac{2}{3})\sqrt{3}+\sqrt{(\sqrt{4}-\sqrt{3})}$

$-(\tfrac{4}{2})\sqrt{2}+3(\tfrac{2}{3})\sqrt{3}-\sqrt{(\sqrt{4}-\sqrt{3})}$

$+(\tfrac{7}{2})\sqrt{2}-5(\tfrac{2}{3})\sqrt{3}+2\sqrt{(\sqrt{4}-\sqrt{3})}$

$-[(\tfrac{11}{2})\sqrt{2}-8(\tfrac{2}{3})\sqrt{3}+3\sqrt{(\sqrt{4}-\sqrt{3})}]$

$+[(\tfrac{18}{2})\sqrt{2}-13(\tfrac{2}{3})\sqrt{3}+5\sqrt{(\sqrt{4}-\sqrt{3})}]$

$-[(\tfrac{29}{2})\sqrt{2}-21(\tfrac{2}{3})\sqrt{3}+8\sqrt{(\sqrt{4}-\sqrt{3})}]$

$5[(\tfrac{3}{2})+\sqrt{4}]\sqrt{2}-34(\tfrac{2}{3})\sqrt{3}+13\sqrt{(\sqrt{4}-\sqrt{3})}$

Table II. cont'd …Generalizations of the Fibonacci Number series of sectionings concluded.

purple: ever-increasing:	ever-decreasing	blue: ever-increasing:	ever-decreasing:
{S:} $\sqrt{4}-\frac{1}{2}\sqrt{6}$	{T:} $\sqrt{4}$	{S:} $\sqrt{(\sqrt{4}-\sqrt{3})}$	{T:} $\sqrt{2}$
(T:} $\sqrt{4}$	{S:} $\sqrt{4}-\frac{1}{2}\sqrt{6}$	(T:} $\sqrt{2}$	{S:} $\sqrt{(\sqrt{4}-\sqrt{3})}$
——————:	——————:	——————:	——————:
$2\sqrt{4}-\frac{1}{2}\sqrt{6}$	$+\frac{1}{2}\sqrt{6}$	$\sqrt{2}+\sqrt{(\sqrt{4}-\sqrt{3})}$	$\sqrt{2}-\sqrt{(\sqrt{4}-\sqrt{3})}$
$3\sqrt{4}-\frac{1}{2}\sqrt{6}$	$\sqrt{4}-2(\frac{1}{2}\sqrt{6})$	$2\sqrt{2}+\sqrt{(\sqrt{4}-\sqrt{3})}$	$-\sqrt{2}+2\sqrt{(\sqrt{4}-\sqrt{3})}$
$5\sqrt{4}-2(\frac{1}{2}\sqrt{6})$	$-\sqrt{4}+3(\frac{1}{2}\sqrt{6})$	$3\sqrt{3}+2\sqrt{(\sqrt{4}-\sqrt{3})}$	$+2\sqrt{2}-3\sqrt{(\sqrt{4}-\sqrt{3})}$
$8\sqrt{4}-3(\frac{1}{2}\sqrt{6})$	$+2\sqrt{4}-5(\frac{1}{2}\sqrt{6})$	$5\sqrt{3}+3\sqrt{(\sqrt{4}-\sqrt{3})}$	$-3\sqrt{2}+5\sqrt{(\sqrt{4}-\sqrt{3})}$
$13\sqrt{4}-5(\frac{1}{2}\sqrt{6})$	$-[3\sqrt{4}-8(\frac{1}{2}\sqrt{6})]$	$8\sqrt{3}+5\sqrt{(\sqrt{4}-\sqrt{3})}$	$+[5\sqrt{2}-8\sqrt{(\sqrt{4}-\sqrt{3})}]$
$21\sqrt{4}-8(\frac{1}{2}\sqrt{6})$	$+[5\sqrt{4}-13(\frac{1}{2}\sqrt{6})]$	$13\sqrt{3}+8\sqrt{(\sqrt{4}-\sqrt{3})}$	$-[8\sqrt{2}-13\sqrt{(\sqrt{4}-\sqrt{3})}]$
$34\sqrt{4}-13(\frac{1}{2}\sqrt{6})$	$-[8\sqrt{4}-21(\frac{1}{2}\sqrt{6})]$	$21\sqrt{3}+13\sqrt{(\sqrt{4}-\sqrt{3})}$	$+[13\sqrt{2}-21\sqrt{(\sqrt{4}-\sqrt{3})}]$
$55\sqrt{4}-21(\frac{1}{2}\sqrt{6})$	$+[13\sqrt{4}-34(\frac{1}{2}\sqrt{6})]$	$34\sqrt{3}+21\sqrt{(\sqrt{4}-\sqrt{3})}$	$-[21\sqrt{2}-34\sqrt{(\sqrt{4}-\sqrt{3})}]$
…	…	…	…

Table III.

green:

ever-increasing:

{S:} [½-(⅓)]√3

{T:} √2-(⅓)√3

———————:

√2+[½-2(⅓)]√3

2√2+[½-3(⅓)]√3

3√2+[2(½)-5(⅓)]√3

5√2+[3(½)-8(⅓)]√3

8√2+[5(½)-13(⅓)]√3

13√2+[8(½)-21(⅓)]√3

…

ever-decreasing:

{T:} √2-(⅓)√3

{S:} [½-(⅓)]√3

———————:

{L:} √2-[½]√3

-√2+[2(½)-(⅓)]√3

+2√2-[3(½)-(⅓)]√3

-3√2+[5(½)-2(⅓)]√3

+5√2-[8(½)-3(⅓)]√3

-8√2+[13(½)-5(⅓)]√3

…

---

red:

ever-increasing:

{S:} -½√2+(⅔)√3

{T:} ½√2+√(√4-√3)

———————————:

(⅔)√3+√(√4-√3)

½√2+(⅔)√3+2√(√4-√3)

½√2+2(⅔)√3+3√(√4-√3)

2(½√2)+3(⅔)√3+5√(√4-√3)

3(½√2)+5(⅔)√3+8√(√4-√3)

5(½√2)+8(⅔)√3+13√(√4-√3)

…

red:

ever-decreasing:

{T:} ½√2+√(√4-√3)

{S:} -½√2+(⅔)√3

———————————:

{L:} 2(½)√2-(⅔)√3+√(√4-√3)

-3(½√2)+2(⅔)√3-√(√4-√3)

+[5(½√2)-3(⅔)√3+2√(√4-√3)]

-[8(½√2)-5(⅔)√3+3√(√4-√3)]

+[13(½√2)-8(⅔)√3+5√(√4-√3)]

-[21(½√2)-13(⅔)√3+8√(√4-√3)]

…

---

orange:

ever-increasing:

{S:} -½√2+(⅔)√3

{L:} +2(½√2)-(⅔)√3+√(√4-√3)

{T:} +½√2+√(√4-√3)

———————:

+3(½√2)-(⅔)√3+2√(√4-√3)

+4(½√2)-(⅔)√3+3 √(√4-√3)

+7(½√2)-2(⅔)√3+5√(√4-√3)

+11(½√2)-3(⅔)√3+8√(√4-√3)

+18(½√2)-5(⅔)√3+13√(√4-√3)

+29(½√2)-8(⅔)√3+21√(√4-√3)

+ 47(½√2)-13(⅔)√3+34√(√4-√3)

+76(½√2)-21(⅔)√3+55√(√4-√3)

…

ever-decreasing:

{T:} ½√2+√(√4-√3)

{L:} +2(½√2)-(⅔)√3+√(√4-√3)

{S:} -(½√2)+(⅔)√3

———————:

+3(½√2)-2(⅔)√3+√(√4-√3)

-[4(½√2)-3(⅔)√3+√(√4-√3)]

+[7(½√2)-5(⅔)√3+2√(√4-√3)]

…

Table III., cont'd

PART II.
Implications of the Fractal Number Architecture for 21st Century Mathematics

Chapter VII. THE CIRCUMFERENCE NUMBER INTEGERS, THE FRACTAL NUMBER INTEGERS AND THEIR IMPLICATIONS FOR 21st CENTURY MATHEMATICS, *QUA APPLIED* MATHEMATICS

Implications of the Theory of the Number *Plane*

1. *All* of current mathematics is an internally consistent intellectual system of logical implications built on the foundation of <u>three</u> *quantitative* <u>and</u> *qualitative* relations of **invariance**.

    The structural foundation for this system is the articulation of a hierarchy of <u>three</u> invariant properties of *THE CIRCLE*:

    A. The '<u>length</u>' 'measure' relation for *any pair* of radii, for <u>any</u> circle;

    B. The length measure ratio for <u>any</u> radius and the 'rim', for <u>any</u> circle;

    C. The '<u>*area*</u>' measure relation between *any* radius and its 'area', for <u>any</u> circle.

    The *architecture*, itself, is a *structure* and a protocol of *processes* which are *constructed* under the strict governance of a total of <u>five</u> rules of 'carpentry'. Each rule is a ***binary point-pair*** **instruction**.

    In *constructing* a system of Mathematics, the meta-mathematician <u>*must*</u> be <u>governed</u> by the five rules in designing the strategic plan to <u>maximize</u> the Teachings in the *Implicate Order Intelligence* embodied in the three invariance implications of *THE CIRCLE*. The definition of the theory of Countable Number is the Procrustean Couch upon which a primitive system of Mathematics is *constructed*.

2. For reasons historical, sociological, psychological, ideological and idiosyncratic, however, the Guardians of the <u>convention</u> for the 'Purity' of our mathematical concept, 'Number', have voted a severe <u>minimization</u> of the implications of A, B, and C for the technical definition of 'Number' and 'Arithmetic'.

    The Epistemological restrictions demanded as cost, by our fastidious <u>minimization</u> [a 2300 year-old inheritance!] of the concept, 'Number', shown in Chapter I, is indicated in the demonstrations of Chapters II-VI. The modern mathematician's continued arbitrary use of Occam's Razor on the mathematical concept, 'Number', has apodized the implications of 'Euclid's' *<u>plane</u>* geometry as the **mother** of modern mathematics:

    Their *ab inito* surgery excludes <u>nearly</u> *<u>all</u>* 'numbers' from the <u>precise</u>

architecture of mathematics, leaving *the least **intelligent*** system of 'number' [and, a *derivative* sub-system, moreover] as the foundation for 'Logic', Set Theory, Arithmetic, Algebra, and 'Analysis'.

Although this *arbitrarily* thus *elected* and *privileged* special case of number does have, also, a planar representation, its linear representation was elected as the *primitive* 'self-replicative, to scale, calibration for the foundation of Arithmetic and Mathematics,

the 'number' <u>line</u>.

3. The Chapters I-VI establish, as a Theorem of [The *Rest* of] 'Euclid', the definition of the number *plane*! The additional **spatial dimension** for the epistemological universe for 'number' definition *permits* precise mathematical expression of the (*planar*) hierarchy of the (infinitely more intelligent) *transcendental* and the *trans-'rational'*, and non-linear **degrees of freedom** for Descartes' (*Analytic Geometry*) Position Vector, and therefore for the Kinematics to enable expression of Newton's and Maxwell's Dynamics.

    With its corollary *particularization* of Set Theory to the restrictions of the elementary theory of Groups having **discrete elements, and with its** *generalization* of Arithmetic, Algebra, Analysis, and *Synthesis*, the theory of the 'number' **plane**, The *Rest* of 'Euclid' delivers an *ab inito* inherently *non*-linear 3rd millennium Mathematics—within which our 2nd millennium Mathematics is **embedded**, as an almost empty, *linear*, special case of ***ARTIFICIAL INTELLIGENCE***.

4. This number *plane* generalization clarifies an august conundrum for modern mathematics. Here is how Eric Temple Bell describes the conundrum:

    *"**The theory of numbers is the last great un**-civilized continent of mathematics. It is split up into **in**-numerable countries, fertile enough among themselves, but all more or less indifferent to one another's welfare **and** without a vestige of a central, intelligent government. If a young Alexander is weeping for a new world to conquer, **it** lies before him. Arithmetic has not had its Descartes, to say nothing of its Newton."*

    – Bell, in Mathematics: Queen and Servant of Science

    The *circle-pair*, with a *shared* radius, is shown to be the *mathematical* **complex** that provides central intelligent government for the last great un-civilized continent of Bell's remark. The geography of the *Transcendental Number* countries, that of the *Trans-'rational' Algebraic Number* countries, and that of the so-called *'Rational' Algebraic Number* country are shown to be hierarchical properties of *THE CIRCLE*, in the *ancient* geometry *plane* of 'Euclid'.

5. Let us consider the relevance of a remark from Gauss' dissertation on the *Implicate Order* role of the (*transcendental*) 'rim' of *THE CIRCLE*:

"*335. Among the splendid developments contributed by modern mathematicians, the theory of circular functions without doubt holds a most important place. We often have occasion in a variety of contexts to refer to this remarkable type of quantity,* **and there is no part of general mathematics that does** **not** **depend on it in some fashion.** *Since the most brilliant mathematicians by their industry and shrewdness have built* ***it*** *into an extensive discipline,* **one would hardly expect any part of the theory,** **let** **alone** **an** **elementary** **part****, could be significantly expanded. I refer to the theory of trigonometric functions corresponding to** *arcs that are commensurable with* *the* *circumference**, i.e.,* **the theory of regular polygons. Only a small part of this theory has been developed so far***, as the present section will make clear.* **The reader might be surprised** *to find a discussion of this subject in the present work which deals with a discipline so unrelated;* **but the treatment itself will make it abundantly clear that there is** **an** **intimate** **connection** *between this subject and* **higher arithmetic.**"

– from: …first article, Section VII, of Gauss' *Disquisitiones Arithmeticae* Jay R. Goldman, *The Queen of Mathematics: A Historically Motivated Guide to Number Theory*. A. K. Peters, Wellesley, Mass (1998), p. 204.[13]

The pair of concentric circle families gives the doubly-infinite table of fractal number integer set of Trigonometric Ratios.

The Cyclotomy of the Three-circle pair, the Four-circle pair and the Ten-circle pair in Chapter V calibrates the Euclidean *plane* into Orthogonal Pairs of *FRACTAL NUMBER INTEGER* Spatial Frequency Systems. The symmetry properties of this set of three *regular polygons* tile the plane.

Implications from Planar Arithmetic

1. Planar Addition Rule; scalar, line rule as a special case.
2. Precise 'fractal number' interpolation of the 'rational' number line calibrations.
3. Positional notation calibration of the Euclidean plane, in powers of the three *plane area-tiling* fractal numbers.
4. With the definition for the magnitude of the reference interval as defined in relation [1],

$$AB \equiv \sqrt{1},$$

the unconventional step was made into the domain of so-called underline{*ir*}-rational 'number'.

A more general system of steps into this domain [where *most* of the '*numbers*' are *positioned*] is provided by the countably infinite set of choices.

$$AB \equiv [p/q]^{[m/n]} \qquad [1']$$

where [p/q] and [m/n] are rational. The pair of choices (q, n) gives a doubly rich set of selections for the Position Vector's interpolation calibrations.

5.   In addition to the logarithm base number systems, [16], [17], and [18], 'Euclid' suggests also,

$$\{\sqrt{1}, (\tfrac{2}{3})\sqrt{3}, (\tfrac{1}{2})\sqrt{3}\} \qquad [16']$$

Implications from Planar Algebra
1.   Synthetic Algebra

It was seen in Chapter IV that the *solution set* for the algebraic equation representation, [14], of the *complex* of the transcendental number integer *pairs* comprises the doubly infinite set of fractal number decompositions of the fractal number integers collected in APPENDIX III.

The *complex* of point patterns, [14], was derived as *a symmetry-breaking* development of Figure 4—the *complete* implication of the reference point-pair, [A, B].

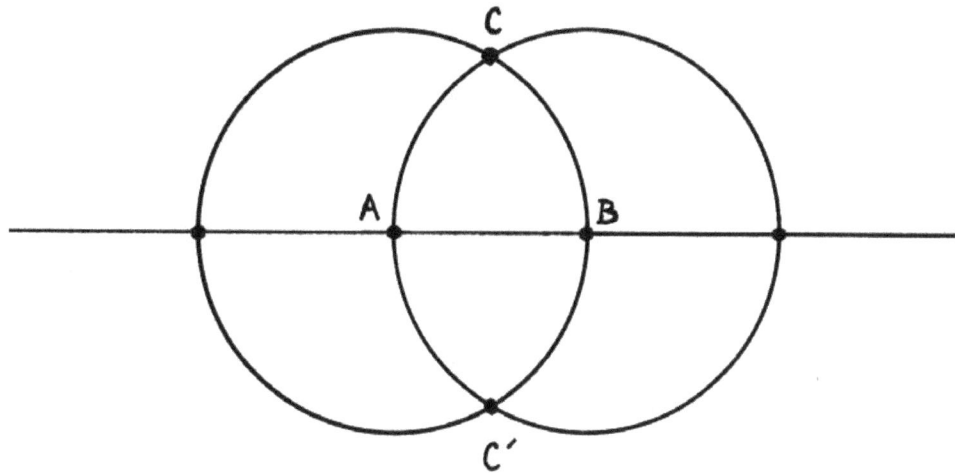

Figure 4

This '...*Rest* of 'Euclid'...', Figure 4, is readily represented by the three 'seed' algebra equations of [14], namely,

$$[x + \tfrac{1}{2}AB]^2 + [y]^2 = [AB]^2$$
$$[x - \tfrac{1}{2}AB]^2 - [y]^2 = [AB]^2$$
$$y = 0 \qquad [16+3]$$

Together with the green line, x = 0 {or the green line x = -½AB, or the green line x = +AB}, the *Algebra* of [16] calibrates the plane with the area positional notations of Figure 22 and Figure 23.

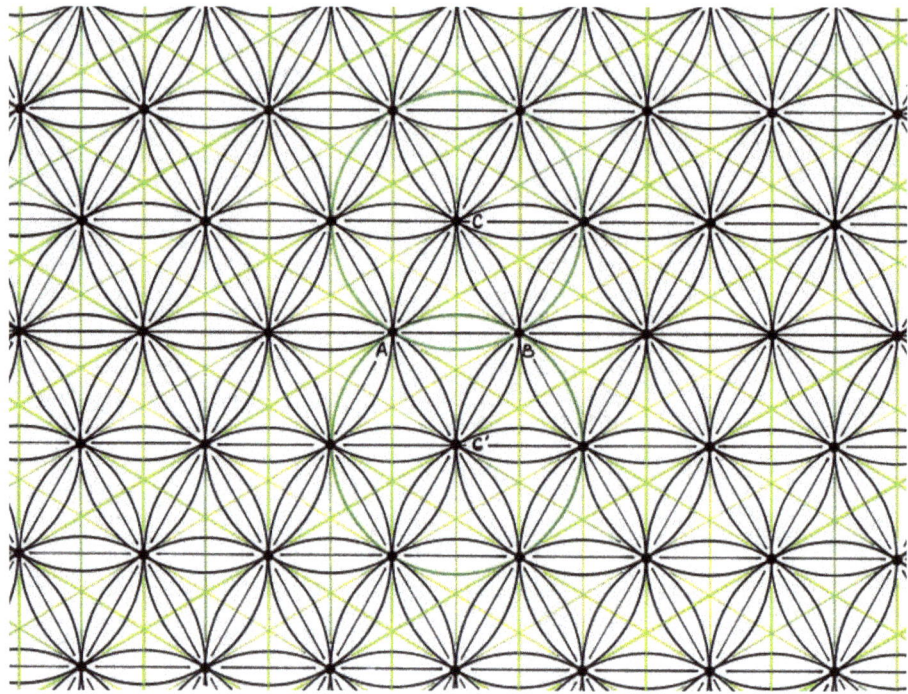

Figure 22. Three ⊥ pairs of √3 Spatial Frequency orientations, the fundamental period.

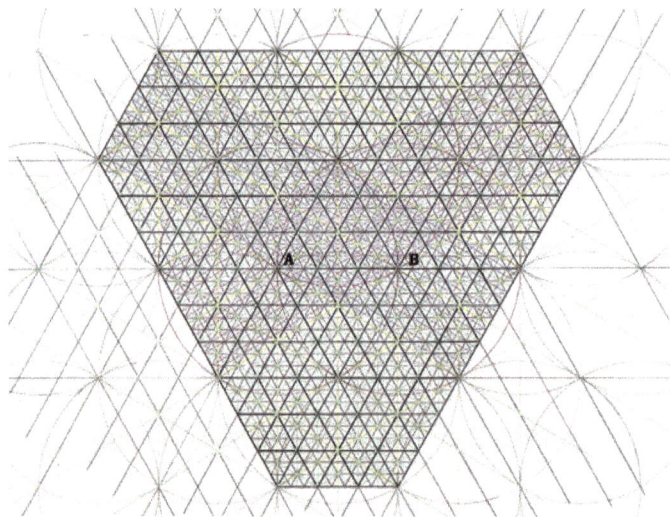

Figure 23. Original vintage drawing by Robert L. Powell, Sr. (See larger image on p. 66.)

It might be expected, further, that the *solution set* for the algebraic equation representation of other transcendental number integer-pair *complexes* do likewise expose the *architecture* of other patterned combinatorial compositions of fractal numbers.

Such candidates for this expectation are the equations for the circle-sets with radii:

$$(\tfrac{2}{3})\sqrt{3}, \sqrt{1}, \tfrac{1}{2}\sqrt{3}; \qquad \sqrt{2}, \sqrt{1}, \tfrac{1}{2}\sqrt{2}; \qquad \tfrac{1}{2}[\sqrt{5}+\sqrt{1}], \sqrt{1}, \tfrac{1}{2}[\sqrt{5}-\sqrt{1}]. \qquad [17]$$

In any case, it is demonstrated that algorithms which strategically examine the results of *symmetry-breaking* the CIRCLE RIM lead to the *synthesizing* of systems of canonical *Algebraic equations, and the complex of point-patterns* which comprise *the complete solution set* for the systems.

2. Synthetic Graph Theory

It is the fashion of Graph Theory to concern with collections of point-pairs and the non-trivial networks of connecting line-segments: "DOTS AND LINES". Little consideration is given to the utilitarian role of THE CIRCLE RIM as an Implicate Order *Intelligence* governing the gestalt efficiency, elegance, and appropriateness of the network.

Richard J. Trudeau reminds us of the Explicate Order foundation in Pure Mathematics, for Graph Theory.[14]

In his Chapter 8, "Platonic Graphs", Trudeau gives the *subtle* teaching of the role of the Theory of Finite Groups in the context of Gauss' Cyclotomy, as an organizing procedure for both *synthesizing* and *analyzing* intelligent patterns of 'dot-pair connected lines'.

The positional notation planar fractal element *plane-filling* powers of [17] give both *arithmetic* and *algebra* to the parallel-organized patterns of self-replicative embedments of patterns of Graph networks within patterns of Graph networks, within patterns of Graph networks, within…Countably many.

3. Clifford (Geometric) Algebras

Maxwell saw fit to use a non-scalar Algebra to organize the Faraday developments. After his death, a less cumbersome branch of Gibbs' non-scalar Algebras became the fashion, custom, and catechism for Dynamics.

Under the auspices of the Canadian Association of Physics, a school of physicists, mathematicians and engineers have been re-examining the geometric structure of the William Kingdon Clifford approach to constructing 'Algebra' on the foundation of 'Euclid'.[15]

Following the Clifford Product of Vectors, back through Grassmann, Hamilton, Rodrigues, Gauss, and others, they report that:

"many concepts in physics are clarified, united, and extended in *new* and *sometimes surprising* directions. In particular, (the approach) eliminates the formal gaps that traditionally separate classical, quantum, and relativistic physics. It thereby makes the study of physics more efficient and the research more penetrating, and it suggests resolutions to a major physics problem of the 20th century, namely how to unite quantum theory and gravity."

It is inconceivable that the Clifford Algebras can serve the tasks of the CAM's school without explicit use of the subtle Kinematics and the (transcendental integer number *governed*) planar Symmetry Properties of the fractal number Position Vector.[16]

4. Implications for Linear Analysis

Historically, the development of 'algebra' was influenced by the study of convergence properties of well-behaved infinite series.[17] A pregnant concern was analysis and/or synthesis of well-behaved and ill-behaved 'functions' of 'x', having the *architecture* [referred to the number *line*]:

$$f(x) = a_0 + a_1x + a_2x^2 + a_3x^3 + \dots + a_nx^n + a_{(n+1)}x^{(n+1)} + \dots \ , \qquad [18]$$

where the a_n 'coefficients' are precisely calculable/*computable* arithmetic numbers.

The analysis/synthesis restricts attention to 'small' values of 'x', i.e., 'x' 'in the neighborhood of' 'x_0' ≡ '0'.

To move the focus of the 'neighborhood' along the number <u>line</u>, to x ≡ ± x_0:

$$f(x \pm x_0) = b_0 + b_1(x \pm x_0) + b_2(x \pm x_0)^2 + b_3(x \pm x_0)^3 + \dots + b_n(x \pm x_0)^n + \dots \ . \quad [18a]$$

The fundamental 'Game' of analysis/synthesis in Algebra is to completely 'factor' such a 'polynomial', for the values of 'x' that are 'sufficiently near' to the neighborhood's focus.

A crucial analytical property of the polynomials is in the questions:

Does the set of '*coefficients*' give a <u>unique</u> representation of the 'function', over the neighborhood? Is each coefficient <u>single-valued</u>, over the neighborhood?

These questions intimately connect 'Algebra' and 'Analysis'—i.e., Algebra and Differential & Integral Calculus.

Modern analysis is careful to identify, and take special cautions with 'functions' with multiple values in the coefficients for its *unique* power series expansion. The 'pathology' of identifiable 'violations' of *Analyticity* is a subject met with enthusiasm.[18,19]

In the case of mathematical representation of structure and process, referred to the **number *plane*,** however, the notion 'function'—as expressed in the form of [18], requires re-thinking its Analytic Geometry meaning.

Now, the 'variable', 'x', is not restricted to 'the number <u>line</u>.' The range of its position, **x**, is the plane. The position *change* has two 'degrees of '*freedom*'.

A richer protocol of *meaning*, then has to be constructed—for some things the likes of:

$$f(\mathbf{x}), \text{ and } f(\mathbf{x} \pm \mathbf{x}_0). \qquad [19a]$$

The necessity for an appropriate re-expression of the Analytic Geometry meaning of 'function', as embodied in [18], has been 'handled' by mathematicians in two encounters.

In 'Vector' *Analysis*, where the '*function*', itself, as well as the planar *position*, **x**, for which the *function's* value(s) is/are specified, <u>also</u> requires multiple 'degrees of freedom' for expression, the notation re-formulation expression is:

$$\mathbf{f(x)}, \text{ and } \mathbf{f(x \pm x_0)} \qquad [19b]$$

The protocol for the other previous encounter is constructed within certain constraints strategically intended to *preserve* certain attractive <u>*analytical*</u> properties exploited in the simpler *architecture* of the number <u>*line*</u> protocol of 'function' theory.[20] A notational representation for this structure re-formulation complexity is:

$$f(x + iy) = u(x, y) + iv(x, y) = G(u + iv) \qquad [19c]$$

This 'theory of the *analytical function* of a 'complex' variable <u>***specializes***</u> the function f (u, v), in the (u, v) plane, to domains of position for which its calculus properties are either 'preserved' or 'accounted for'.

That is to say, the theory relates a (two-dimensional, *structured*) function, G(u, v), in a u-v <u>*plane*</u>, to a condition-restricted domain of *positions* in the *structured* x-y <u>*plane*</u>.

In the case of Algebra and Analysis, referred to the number <u>*plane*</u>, both the theory of the analytical function of the complex variable and the theory of vector analysis are touched.

In the number plane, as developed above, the geometric *objects* that invite the gestalt notion, <u>*function*</u>, are, *themselves*, (two-dimensional) <u>numbers,</u> or arithmetic and algebraic <u>combinatorials</u> of planar <u>numbers.</u>

The arithmetic and algebraic '*functions*' are such planar geometric objects as directed line segments, directed arcs of circles, directed area elements of circles, directed area elements of polygons.

The planar *independent* variable domain for the Theory of the Analytical Function of the Complex Variable has become the planar domain in which to <u>*synthesize*</u> arithmetic and algebraic combinatorial <u>***computations***</u> for the transcendental number integer, p, and the trans-rational number integers, $\sqrt{1}$, $\sqrt{2}$, $\sqrt{3}$, ...

Not only is the study of synthesis/analysis of these functions relieved of the Procrustean Couch requirement of analyticity, indeed, the three impositions of the Quantum Rules of Gauss' Cyclotomy introduce the three regular <u>violations</u> of *analyticity* to provide the eigen-numbers for universal *fractal* self-replications, to-scale.

The countably infinite combinatorial degrees of freedom of position vector *trajectory* afforded by the trans-rational numbers....

5.   Fractal Number Spatial Frequency Spaces.

The establishment of the *second* one of the reference Binary point-pair [A, B] immediately determines the three systems of orthogonal pairs of spatial frequency calibrations for the plane. These systems are exemplified in Figures 23, 26, and 28.

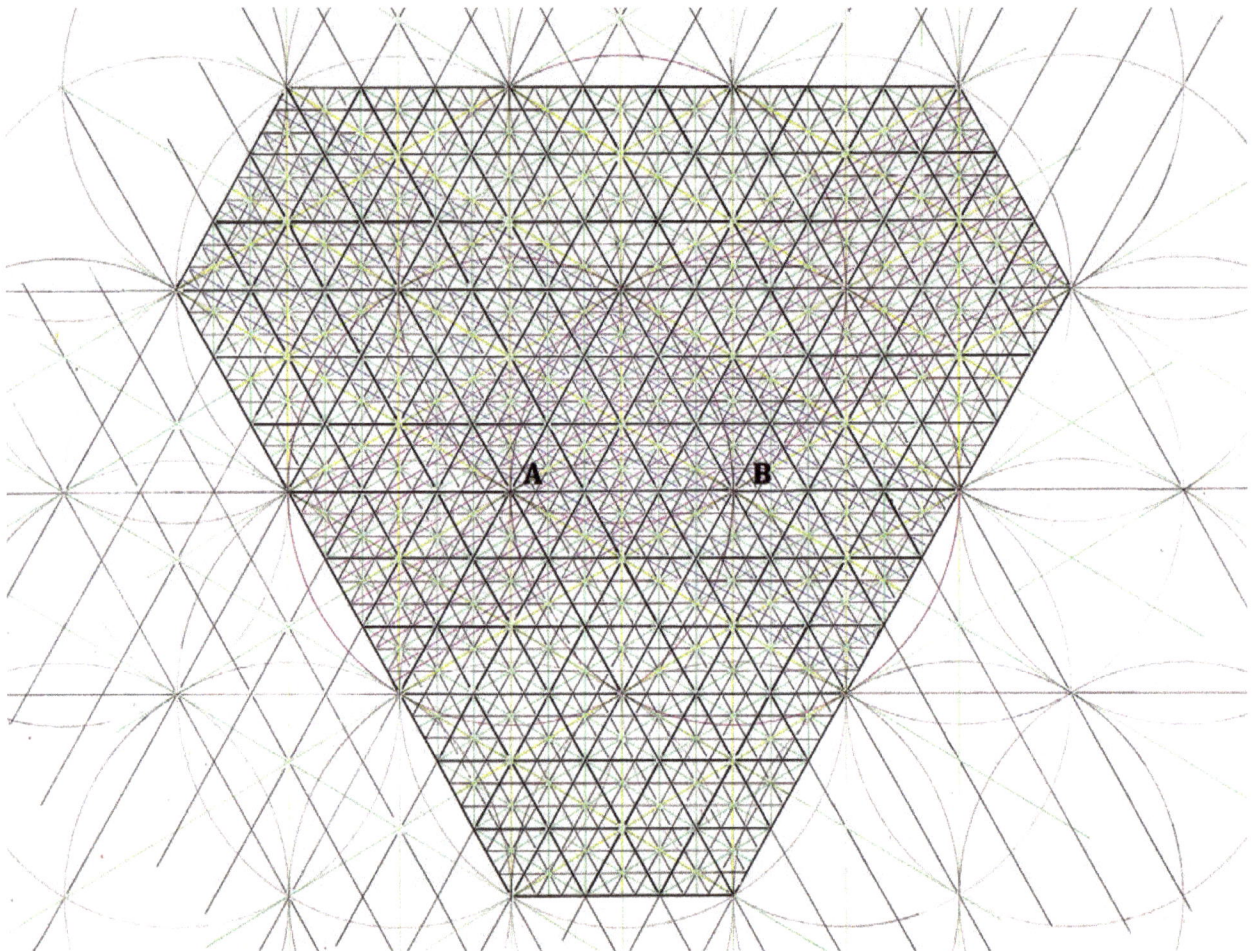

Figure 23. Original vintage drawing by Robert L. Powell, Sr.

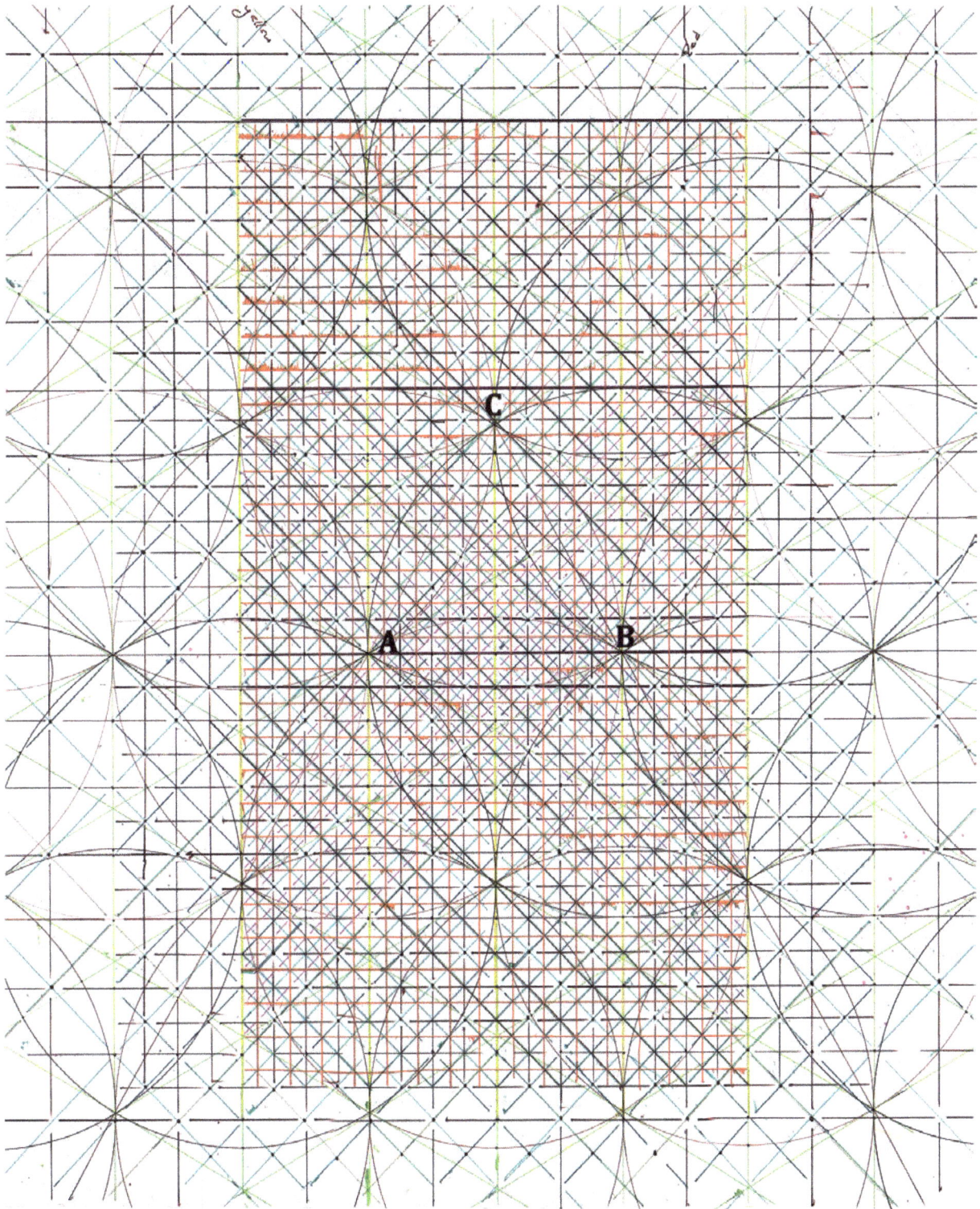

Figure 26. The *planar* space-filling √2 number fractal co-ordinate system. Original vintage drawing by Robert L. Powell, Sr.

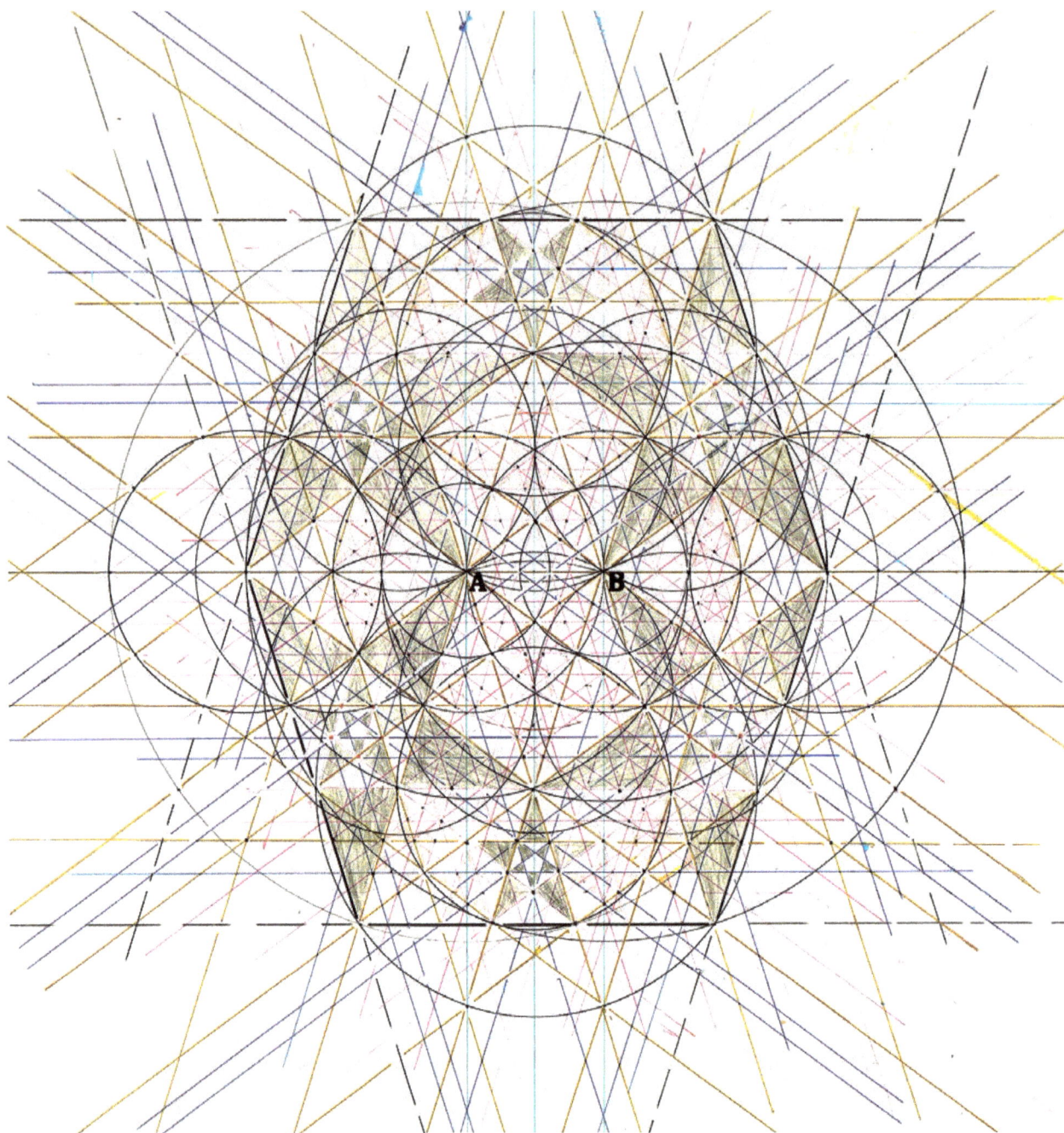

Figure 28a. Original vintage drawing by Robert L. Powell, Sr.

13 Jay R. Goldman, *The Queen of Mathematics: A Historically Motivated Guide to Number Theory*. A. K. Peters, Wellesley, Mass (1998), Chapter 14.

14 Richard J. Trudeau. *Introduction to Graph Theory*. Dover (1993). Chapter 1; Chapter 8.

15 William E. Baylis, ed. *Clifford (Geometric) Algebras with Applications in Physics, Mathematics, and Engineering*. Birkhauser Boston (1996)

16 O. Bottema and B. Roth, *Theoretical Kinematics*. Dover (1990)

17 Shaughan Lavine. *Understanding the Infinite*, Harvard U. Press (1994)

18 David Berlinski. *A Tour of the Calculus*, Vintage Books. New York (1997)

19 Hermann Weyl. *The Continuum, A Critical Examination of the Foundation of Analysis*. Dover. New York (1994)

20 David Berlinski. *A Tour of the Calculus*, Vintage Books. New York (1997)

Chapter VIII. **THE TRANSCENDENTAL NUMBER INTEGERS, AND THEIR FRACTAL NUMBER INTEGERS FOR 21ST CENTURY MATHEMATICS, *QUA APPLIED* MATHEMATICS**

As preparation to recognize and examine specific utilitarian scientific and technological value of The Knowledge of The Number Plane, compared to the mere Knowledge embodied in the *complexity* of The Number Line's calibrations, let us review the previous chapter from the point of view of Descartes' notion, the Position Vector.

With reference to the A point, the B point or the O point, in Figure 9, say:

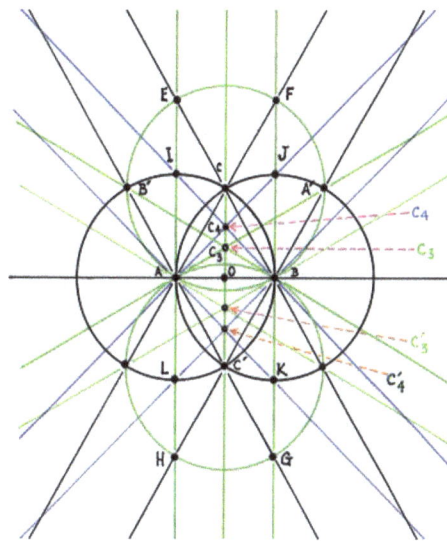

Figure 9.

1. <u>Every</u> point created canonically in the plane is a *computed location*.
2. <u>Every</u> canonically constructed circumference segment and <u>every</u> rule-constructed line segment is, and *only* is, the *potential location* for a *computed point*—the *two*-dimensional, planar, position of a *zero*-dimensional (point) universe.
3. Every canonically created line segment is an (algebraic) *number-computed* interval—a *two*-dimensionally *directed* <u>interval</u>, of a *one*-dimensional continuum universe.
4. Every canonically constructed circumference arc is a *transcendental number-computed* area sector, of a *two*-dimensional *transcendental number* area continuum universe.

The coherent algorithm of canonical *computations* then <u>teaches</u> the strategic introduction of the **complex** of a countably many **degrees of freedom** within the space of a **two-dimensional**, planar position of a <u>zero</u>-dimensional point universe.

The mathematical complexity of the *degrees of freedom* of the machine language's Position Vector, and the in principle <u>zero</u>-dimensional universe of its sequentially created set of <u>points</u> are inherent liberations of the art of machine computation hardware and software, from the <u>one</u>-dimensional constraint of the von Neumann-Turing architecture.

(1) **<u>COMPUTER SCIENCE: The Hardware</u>**
The von Neumann-Turing <u>architecture</u> for Artificial Intelligence computing machines confines the *intelligence* of existing *state of the art* super-computing machine complexes to combinatorial *computations* coordinated on to only a single *'rational' number* <u>line</u>.

To give a *machine*-computed out-put even the <u>*simulated*</u> **appearance** of an <u>inherently</u> parallel-organized <u>*computed*</u> result is, today, the cutting-edge challenge for the super-computer machine designer.

The very last public remark of von Neumann, printed posthumously, refers explicitly to the as-yet still not generally recognized Epistemological limitation of the <u>mathematical</u> language, which the computation <u>machine</u> language faithfully *represents*.

In commenting on our <u>mathematical</u> language as a capacity for *faithful* representation of structures and processes of *NATURAL Intelligence*, he concludes[21]:

"...We have now accumulated sufficient evidence to see [that] whatever the language the central nervous system is using, it is characterized by less logical and arithmetical depth than we are normally used to...Consequently, there exists here different logical structures from the ones we are ordinarily used to in logic and mathematics...Thus logics and mathematics in the central nervous system, when viewed as languages, **must structurally be <u>essentially</u> different from those languages to which our common experience refers.***"*

***"It also ought to be noted that the language here involved may well correspond to a short code...rather than to a complete code: when <u>we</u> talk mathematics <u>we</u> may be discussing** a <u>secondary</u> language, built on the <u>primary</u> language <u>truly</u> used by the central nervous system. Thus the outward form of <u>our</u> mathematics are not absolutely relevant from the point of view of evaluating what the mathematical or logical language <u>truly</u> used by the central nervous system <u>**is**</u>, <u>it</u> cannot fail to differ considerably from <u>what</u> <u>we</u> <u>consciously</u> and <u>explicitly</u> <u>consider</u> as <u>**mathematics**</u>."*[22]

Expanding on von Neumann's remark, we assert that the computation machine language taught by the foregoing Theory of the Number Plane is the **complete code**, the <u>primary</u> mathematical language <u>truly</u> used by <u>**NATURE**</u>; that "what we consciously and explicitly consider as *mathematics*" is a <u>secondary</u> language, a *short code*, embedded within the more <u>nature</u>-*like* language of The Pair of Circles, with a Shared Radius.

The machine hardware to store sequentially the accumulated lines of canonical code and to present the parallel-organized patterned computations must be able to *write precisely* and to *present* the transcendental number circumferences and their determined *directed* line segment algebraic numbers.

As the *visual* diagrams in this book Teach, such a machine language is readily Learned, Designed, and Manufactured. Without such a line of manufacture, the mathematics of the 3rd millennium must be computed by hand, as of old.

(2) **COMPUTER SCIENCE: The Software**

The von Neumann-Turing *architecture*-produced presentations are one-dimensional scannings contrived so as to *simulate* patterned self-replications of one or more two-dimensional symmetry elements.

Such accomplishments entail a *tour de force* of fragile, expensive, and esoteric accumulations of software code of 'rational' number digit instruction statements.[23, 24] The necessary employment of the self-replication properties of Group Theory/symmetry is *ad hoc*, whimsical, and smacks of erudite mathematical legerdemain.

The programs of Symmetry-breaking in Part I of this book provide the syllabus for the systematic development of a portfolio of one-dimensional, 'rational' number *'short code'* software simulations of the *'complete code'* representations of the transcendental number and trans-rational number mathematical computation structures.

In particular, for example, the more foundational mathematical content of Graph Theory[25, 26], when informed by the fractal number positional notation calibrations for the plane, gives *system* to the tasks of *analyzing* and *synthesizing* tactical networks of Intelligent Agent *complexes*.[27, 28, 29]

(3) **COMPUTER SCIENCE: The Humanware**

The **mathematical *complexity*** of computations using the hierarchy of planar numbers both requires and awakens the anagogic attention of the 'computer—*ex machina*':

Both the *Implicate Order* role[30] and the *Explicate Order* role of the transcendental number circumferences *"differ considerably from what we consciously and explicitly consider as* **mathematics**.*"*

The plane numbers theory permits the human to use the machine to assemble cybernetic[31] models of structure and process which require the precision of transcendental number and *trans*-'rational' number.

(4) INCIPIENT METHODS: *Digital Analysis-Synthesis of Structure-Process*

During the last half of the 20th century, the upright bipedal scientific creature has sought, consciously, to capture the *mathematical* <u>representation</u> of the Implicate Order *complexity* of <u>perceived</u> structure and process.

This Herculean investment in the <u>*scientific*</u> Analysis-Synthesis of Structure-Process has its *mathematical <u>basis</u>* in **the dumbest <u>conceivable</u>** set of computable planar *<u>point</u>-pattern* intervals:

> the <u>point</u>-pattern that characterizes the pure <u>*one*</u>-dimensional step function, with <u>*one*</u> replication interval, [A, B].

This **legerdemain** produces a <u>simulation</u> that 'looks like' something two- or three-dimensional.

The simulations contrived by the von Neumann-Turing architecture Hardware-Software produce models of scientific structure-process in which the mathematical complexity of **most of 'Euclid'** has been *apodized*.

During the last quarter of the 20th century, some mathematically intuitive *non*-mathematician scientists have been led to contrive mathematical methods of digital analysis-synthesis which conduct more intelligent theoretical and empirical Inquiry than that afforded by the rational number system.

(A) One such construction is The Renormalization Group Method.[(32)] The method makes careful *ad hoc* use of the Explicate Order properties of the fractal number symmetry invariants. The hierarchy of planar fractal spatial frequencies provided by the fractal number lattices given in [16]-[18] make for the Implicate Order Euclidean foundation for the method of digital number analysis-synthesis.

(B) Another method that refines the art of digital analysis-synthesis of structure-process is the mathematical technique, Wavelet Analysis.[(33, 34)] The algebraic number *integers* of the fractal number plane's hierarchical *positional notation* tiling provides a transparent geometric foundation for a canonical algebra for the wavelet analysis-synthesis task.

(C) A third emerging method of digital synthesis-analysis can immediately benefit from the position notation provided by the fractal number integer calibrations of the plane is the study of cyclostationary processes.[(35)] A cyclostationary process is one in which one or more statistical periodicities is present.

The fractal number Cyclotomy values for the Three-, Four-, and Ten-circles—[16], [17], and [18]—might assist the engineers and scientists in the growing awareness that cyclostationary processes are, "in many ways, much more than a trivial variation on stationary processes and do therefore merit their attention to develop and refine the theory of these processes."[(36)]

21 John von Neumann. *The Computer and The Brain*, Yale U. Press (1958), pp 80-82.

22 Ibid.

23 Ibid.

24 Ibid.

25 Ibid.

25 Richard J. Trudeau. *Introduction to Graph Theory*. Dover (1993), Chapters 4-8.

26 Gary Chartrand, *Introductory Graph Theory*. Dover (1984)

27 Richard J. Trudeau. *Introduction to Graph Theory*. Dover (1993)

28 Ibid.

29 Ibid.

30 David Bohm. *Fragmentation and Wholeness*, Van Leer Publisher, Jerusalem (1976)

31 Norbert Wiener. *Cybernetics: or Control and Communication in the Animal and the Machine*. Cambridge, Mass. MIT Press (1969)

32 R. J. Creswick, H.A. Farach, C.P. Poole, Jr. *Introduction to Renormalization Group Methods in Physics*. John Wiley Sons, New York (1992).

33 Barbara Burke Hubbard. *The World According to Wavelets: The Story of a Mathematical Technique in the Making*. A. K. Peters, Wellesley Mass (1998).

34 Howard L. Resnikoff, Raymond O. Wells, Jr., *Wavelet Analysis: The Scalable Structure of Information*. Springer, New York (1998).

35 William A. Gardner, editor. *Cyclostationarity – in Communications and Signal Processing*, IEEE Press. IEEE Communications Society, Sponsor. New York (1994)

36 William A. Gardner, editor. *Cyclostationarity – in Communications and Signal Processing*, IEEE Press. IEEE Communications Society, Sponsor. New York (1994), p. 4.

Conclusion

Postscript: An Invitation to take up this ongoing research work and its extensive applications

Here are draft notes by Robert L. Powell, Sr.
for additional sections of this book.

Part III.
Some Implications of Part II
of The *REST* of 'Euclid' for
21st Century Science

Chapter IX.   BIOLOGY OF THE LIVING CELL and CHIRALITY / DRAFT NOTES, RLP, SR.

The fourth state of matter: The Theory of the Number Plane is necessary and sufficient for theoretical models and experimental design parameters for this complexity.

After about three decades into the 20th century, the attention of 'Western' elite scientific Inquiry had become focused on the micro-scale and the cosmo-scale:

What is the stability of the atom, as a structure; as a process?

What is the age of the Cosmos, as a structure; as a cosmos?

Two delightful and intelligible summary reports of progress in this federally funded Inquiry-pair, conducted by the century's Apprentice to the Sorcerer, the atom-cosmos physicist are:

(a) The Morrisons characteristically stunning journey across the universe, inferring the size of 'things', from $[10^{+24}$ meters] 'big' to 'things' as small as $[10^{-16}]$ meters[37]

(b) Brian Greene's breath-taking report of the current alchemical proposition about the relativity of the very fabric of the cosmos.[38]

During the same time still another enclave of alchemists was arousing still another ancient Sorcery Apprenticeship. Vestiges of the discredited heresy, 'vitalism', began to stir again—within as well as, of course, outside the Cathedral of Biology of the 'Living'.

(a) One such intra-cathedra heretic, in search of elusive experimental basis for missing theory, was the marine biologist, Ernest Everett Just.[39] His report was almost stifled by the 'mechanists'.[40]

(b) On the matter of the absence of any as yet undiscovered credible evidence for a subtlety-complexity of 'alive' structure-process, however, the indomitable heretic was Albert Szent-Gyorgyi.[41]

(c) On the matter of the absence of any as yet discovered intra cathedra credible evidence for the subtlety-complexity of some 'vital' structure-process, however, an elegant intra cathedra meditation-cogitation was delivered by Erwin Schrodinger.[42]

The seminal *ex cathedra* heretic on this Inquiry was Albert Szent-Gyorgyi.[43] Szent-Gyorgyi's work[44-49] led him to revive earlier attention, by Sir William Hardy (1931) and Hardy & Bircumshaw (1925), the notion of a fourth state of matter.[50]

(d) Goaded by intuition, a community of researchers influenced by Schrodinger and Szent-Gyorgyi has continued the attempt to develop credible theory and sufficiently delicate experimental technique for the level of non-invasion, indeed non-destruction, demanded for meaningful measurements on biologically significant complexes of matter.

The meaningful experimental physics measurements engage the novel subtlety which characterizes the physics measurements vital structure-process.

37 Philip Morrison, Phylis Morrison, the Office of Charles and Ray Eames, *Powers of Ten: about the Relative Size of Things in the Universe and the Effect of Adding another Zero.* Scientific American Library. New York (1994).
38 Brian Greene. *The Fabric of the Cosmos: Space, Time, and the Texture of Reality.* Alfred Knopf (2005)
39 Ernest Everett Just. *The Biology of the Cell Surface.* P. Blakistons Sons & Co. Inc., Philadelphia. (1939).
40 Kenneth R, Manning. *Black Apollo of Science: The Life of Ernest Everett Just.* Oxford U. Press (1985)
41 Erwin Schrodinger. *What Is Life?*
42 Ibid.
43 Ralph W. Moss. Foreword by Studs Terkel. *Free Radical: Albert Szent-Gyorgyi and the Battle over Vitamin C.* Paragon House Publishers, New York (1988).
44 Albert Szent-Gyorgyi. *Nature of Life: A Study on Muscle.* Academic Press, New York (1948)
45 ____. *The Chemistry of Muscular Contraction,* Academic Press, New York (1947)
46 ____. *Chemistry of Muscular Contraction.* 2nd edition, revised and enlarged, Academic Press, New York (1951).
47 ____. *Bioenergetics,* Academic Press, New York (1957).
48 ____. *Introduction to a Submolecular Biology.* Academic Press. New York (1960)
49 ____. *The Living State with Observations on Cancer.* Academic Press. (1972) New York (1972).
50 ____. *The Living State with Observations on Cancer.* Academic Press. (1972), p. 13

Chapter X.   **NANOTECHNOLOGY and BIOMIMICRY / DRAFT**

Chapter XI.   **PHYSICS / DRAFT**
Inherently non-linear classical kinematics & dynamic
precise mathematical representations for String Theory

Part IV.
Some Implications of Part II of The *REST* of 'Euclid' as Paradigm for a Third Millennium Politics of Ecology

Chapter XII.   **A THIRD MILLENNIUM POLITICS OF ECOLOGY / DRAFT**

Appendices

Appendix I.
JACOB'S LADDER: THE NUMBER PATTERNS

JACOB'S LADDER, 1507. November 15, 2003
(p. 1 of 4)

$(\sqrt{n})2$ 'x'=½√1:	'x'=2(½√1):	'x'=3(½√1):	'x'=4(½√1):
1 [½√1, ½√3]			
2 [" , ½√7]	[2(½√1), √1]		
3 [" , ½√11]	[" , √2]	[3(½√1), ½√3]	
4 [" , ½√15]	[" , √3]	[" , ½√7]	
5 [" , ½√19]	[" , √4]	[" , ½√11]	[4(½√1), √1]
6 [½√1, ½√23]	[" , √5]	[" , ½√15]	[" , √2]
7 [" , ½√27]	[2(½√1), √6]	[" , ½√19]	[" , √3]
8 [" , ½√31]	[" , √7]	[3(½√1), ½√23]	[" , √4]
9 [" , ½√35]	[" , √8]	[" , ½√27]	[" , √5]
10 [" , ½√39]	[" , √9]	[" , ½√31]	[4(½√1), √6]
11 [½√1, ½√43]	[" , √10]	[" , ½√35]	[" , √7]
12 [" , ½√47]	[2(½√1), √11]	[" , ½√39]	[" , √8]
13 [" , ½√51]	[" , √12]	[3(½√1), ½√43]	[" , √9]
14 [" , ½√55]	[" , √13]	[" , ½√47]	[" , √10]
15 [" , ½√59]	[" , √14]	[" , ½√51]	[4(½√1), √11]
16 [½√1, ½√63]	[" , √15]	[" , ½√55]	[" , √12]
17 [" , ½√67]	[2(½√1), √16]	[" , ½√59]	[" , √13]
18 [" , ½√71]	[" , √17]	[3(½√1), ½√63]	[" , √14]
19 [" , ½√75]	[" , √18]	[" , ½√67]	[" , √15]
20 [" , ½√79]	[" , √19]	[" , ½√71]	[4(½√1), √16]
21 [½√1, ½√83]	[" , √20]	[" , ½√75]	[" , √17]
22 [" , ½√87]	[2(½√1), √21]	[" , ½√79]	[" , √18]
23 [" , ½√91]	[" , √22]	[3(½√1), ½√83]	[" , √19]
24 [" , ½√95]	[" , √23]	[" , ½√87]	[" , √20]
25 [" , ½√99]	[" , √24]	[" , ½√91]	[4(½√1), √21]
…	…	…	…
…	…	…	…
n[½√1, ½√(4n-1)]	[2(½√1), √{n-1)	[3(½√1), ½√(4n-3²)]	[4(½√1), ½√(4n-4²)
…	…	…	…
…	…	…	…

JACOB'S LADDER, 1507. November 15, 2003
(p. 2 of 4)

$(\sqrt{n})\,2$	'x'=5(½√1):	'x'=6(½√1):	'x'=7(½√1):	'x'=8(½√1):
	…			
	…			
7	[5(½√1), ½√3]			
8	[" , ½√7]			
9	[" , ½√11]			
10	[" , ½√15]	[6(½√1), √1]		
11	[" , ½√19]	[" , √2]		
12	[5(½√1), ½√23]	[" , √3]		
13	[" , ½√27]	[" , √4]	[7(½√1), ½√3]	
14	[" , ½√31]	[" , √5]	[" , ½√7]	
15	[" , ½√35]	[6(½√1), √6]	[" , ½√11]	
16	[" , ½√39]	[" , √7]	[" , ½√15]	
17	[5(½√1), ½√43]	[" , √8]	[" , ½√19]	[8(½√1), √1]
18	[" , ½√47]	[" , √9]	[7(½√1), ½√23]	[" , √2]
19	[" , ½√51]	[" , √10]	[" , ½√27]	[" , √3]
20	[" , ½√55]	[6(½√1), √11]	[" , ½√31]	[" , √4]
21	[" , ½√59]	[" , √12]	[" , ½√35]	[" , √5]
22	[5(½√1), ½√63]	[" , √13]	[" , ½√39]	[8(½√1), √6]
23	[" , ½√67]	[" , √14]	[7(½√1), ½√43]	[" , √7]
24	[" , ½√71]	[" , √15]	[" , ½√47]	[" , √8]
25	[" , ½√75]	[6(½√1), √16]	[" , ½√51]	[" , √9]
26	[" , ½√79]	[" , √17]	[" , ½√55]	[" , √10]
27	[5(½√1), ½√83]	[" , √18]	[" , ½√59]	[8(½√1), √11]
28	[" , ½√87]	[" , √19]	[7(½√1), ½√63]	[" , √12]
29	[" , ½√91]	[" , √20]	[" , ½√67]	[" , √13]
30	[" , ½√95]	[6(½√1), √21]	[" , ½√71]	[" , √14]
31	[" , ½√99]	[" , √22]	[" , ½√75]	[" , √15]
32	[5(½√1), ½√103]	[" , √23]	[" , ½√79]	[8(½√1), √16]
33	[" , ½√107]	[" , √24]	[7(½√1), ½√83]	[" , √17]
34	[" , ½√111]	[" , √25]	[" , ½√87]	[" , √18]
35	[" , ½√115]	[6(½√1), √26]	[" , ½√91]	[" , √19]
	…	…	…	…
	…	…	…	…
n	[5(½√1), ½√(4n-5²)]	[6(½√1), ½√(4n-6²)]	[7(½√1), ½√(4n-7²)]	[8(½√1), ½√(4n-8²)]

JACOB'S LADDER, 1507. November 15, 2003
(p. 3 of 4)

(√n)2:	'x' = 9(½√1)	'x' = 10(½√1):	'x' = 11(½√1):	'x' = 12(½√1):
	…			
	…			
21	[9(½√1), ½√3]			
22	[" , ½√7]			
23	[" , ½√11]			
24	[" , ½√15]			
25	[" , ½√19]			
26	[9(½√1), ½√23]	[10(½√1), √1]		
27	[" , ½√27]	[" , √2]		
28	[" , ½√31]	[" , √3]		
29	[" , ½√35]	[" , √4]		
30	[" , ½√39]	[" , √5]		
31	[9(½√1), ½√43]	[!0(½√1), √6]	[11(½√1), ½√3]	
32	[" , ½√47]	[" , √7]	[" , ½√7]	
33	[" , ½√51]	[" , √8]	[" , ½√11]	
34	[" , ½√55]	[" , √9]	[" , ½√15]	
35	[" , ½√59]	[" , √10]	[" , ½√19]	
36	[9(½√1), ½√63]	[10(½√1), √11]	[11(½√1), ½√23]	
37	[" , ½√67]	[" , √12]	[" , ½√27]	[12(½√1), √1]
38	[" , ½√71]	[" , √13]	[" , ½√31]	[" , √2]
39	[" , ½√75]	[" , √14]	[" , ½√35]	[" , √3]
40	[" , ½√79]	[" , √15]	[" , ½√39]	[" , √4]
41	[9(½√1), ½√83]	[10(½√1), √16]	[11(½√1), ½√43]	[" , √5]
42	[" , ½√87]	[" , √17]	[" , ½√47]	[12(½√1), √6]
43	[" , ½√91]	[" , √18]	[" , ½√51]	[" , √7]
44	[" , ½√95]	[" , √19]	[" , ½√55]	[" , √8]
45	[" , ½√99]	[" , √20]	[" , ½√59]	[" , √9]
46	[9(½√1), ½√103]	[10(½√1), √21]	[11(½√1), ½√63]	[" , √10]
47	[" , ½√107]	[" , √22]	[" , ½√67]	[12(½√1), √11]
48	[" , ½√111]	[" , √23]	[" , ½√71]	[" , √12]
49	[" , ½√115]	[" , √24]	[" , ½√75]	[" , √13]
50	[" , ½√119]	[" , √25]	[" , ½√79]	[" , √14]
	…	…	…	…
	…	…	…	…
n	[9(½√1, ½√(4n-9²)]	[10(½√1), ½√(4n-10²)]	[11(½√1), ½√(4n-11²)]	[12(½√1), ½√(4n-12²)]

JACOB'S LADDER, 1507. November 15, 2003
(p. 4 of 4)

$(\sqrt{n})2$:	'x'=13(½√1):	'x'=14(½√1):	'x'=15(½√1):	'x'=16(½√1):
...				
43	[13(½√1), ½√3]			
44	[" , ½√7]			
45	[" , ½√11]			
46	[" , ½√15]			
47	[" , ½√19]			
48	[13(½√1), ½√23]			
49	[" , ½√27]			
50	[" , ½√31]	[14(½√1), √1]		
51	[" , ½√35]	[" , √2]		
52	[" , ½√39]	[" , √3]		
53	[13(½√1), ½√43]	[" , √4]		
54	[" , ½√47]	[" , √5]		
55	[" , ½√51]	[14(½√1), √6]		
56	[" , ½√55]	[" , √7]		
57	[" , ½√59]	[" , √8]	[15(½√1), ½√3]	
58	[13(½√1), ½√63]	[" , √9]	[" , ½√7]	
59	[" , ½√67]	[" , √10]	[" , ½√11]	
60	[" , ½√71]	[14(½√1), √11]	[" , ½√15]	
61	[" , ½√75]	[" , √12]	[" , ½√19]	
62	[" , ½√79]	[" , √13]	[15(½ √1), ½√23]	
63	[13(½√1), ½√83]	[" , √14]	[" , ½√27]	
64	[" , ½√87]	[" , √15]	[" , ½√31]	
65	[" , ½√91]	[14(½√1), √16]	[" , ½√35]	[16(½√1), √1]
66	[" , ½√95]	[" , √17]	[" , ½√39]	[" , √2]
67	[" , ½√99]	[" , √18]	[15(½√1), ½√43]	[" , √3]
68	[13(½√1), ½√103]	[" , √19]	[" , ½√47]	[" , √4]
69	[" , ½√107]	[" , √20]	[" , ½√51]	[" , √5]
70	[" , ½√111]	[14(½√1), √21]	[" , ½√55]	[16(½√1), √6]
	...	...	...	...
	...	...	...	...
n	[13(½√1, ½√(4n-13²)]	[14(½√1), ½√(4n-14²)]	[15(½√1), ½√(4n-15²)]	[16(½√1), ½√(4n-16²)]
	...			

Appendix II.
More Hinnant-Powell Numbers
(p. 1 of 5)

GREEN RAY, ever increasing:

Small segment increment:

S; (0)   + (1) ½ √3 - (1) (1/3) √3
L: (1) √2 - (1) ½ √3
T: (1) √2+ (0)   - (1) (1/3) √3

large segment increment:

(0)   + (1) ½ √3 - (1) (1/3) √3
(1) √2 - (1) ½ √3
(1) √2 + (0)   -   (1/3) √3

....

1.	T:	√2	- (1/3)√3
2.	T+S:	√2 + ½ √3	- 2 (1/3)√3
3.		2 √2 + ½ √3	- 3 (1/3)√3
4.		3 √2 + 2 (½ √3)	- 5 (1/3)√3
5.		5 √2 + 3 (½ √3)	- 8 (1/3)√3
6.		8 √2 + 5 (½ √3)	- 13 (1/3)√3
7.		13 √2 + 8 (½ √3)	- 21 (1/3)√3
8.		21 √2 + 13 (½ √3)	- 34 (1/3)√3

large segment:

T: √2   - (1/3)√3
T+L: 2 √2 - 1 ½ √3 - 1 (1/3)√3
3 √2 - 1 ½ √3 - 2 (1/3)√3
5 √2 - 2 (½ √3) - 3 (1/3)√3
8 √2 - 3 (½ √3) - 5 (1/3)√3
13 √2- 5 (½ √3) - 8(1/3)√3
21 √2 – 8 (½ √3) - 13 (1/3)√3
34 √2 – 13 (½ √3)- 21 (1/3)√3

"   ....
..   ..
n f_n √2 + $f_{(n-1)}$ (½ √3) - $f_{(n+1)}$ (1/3)√3 $f_{(n+1)}$ √2 – $f_{(n-1)}$(1/3)√3 - $f_{(n)}$ (½ √3)

GREEN RAY, ever decreasing:

1. T: [√2   - (1/3)√3]
  - L: -[√2 - ½ √3   ]
  = S: [½ √3 - (1/3)√3]

2.   [√2 - ½ √3   ]
  - [½ √3 - (1/3)√3]
  = [√2 - 2 (½ √3)+ (1/3)√3]

3.   [½ √3 - (1/3)√3]
  -[√2 - 2 (½ √3) + (1/3)√3]
  =[- √2 + 3 (½ √3) - 2 (1/3)√3]

4.   [√2 - 2 (½ √3) + (1/3)√3]
  - [- √2 + 3 (½ √3) - 2 (1/3)√3]
  = [2√2 - 5 (½ √3 + 3 (1/3)√3]

5.   [- √2 + 3 (½ √3) - 2 (1/3)√3]
  -[2√2 - 5 (½ √3) + 3 (1/3)√3]
  =[- 3√2 + 8 (½ √3) - 5 (1/3)√3]

6.   [2√2 - 5 (½ √3) + 3 (1/3)√3]
  - [- 3√2 + 8 (½ √3) - 5(1/3)√3]
  = [5√2 - 13 (½ √3) + 8(1/3)√3]

7.   [- 8√2 +21 (½ √3 - 13 (1/3)√3

8.   [13√2 – 34 (½ √3) + 21(1/3)√3]

9.   [- 21√2 + 55 (½ √3) - 34 (1/3)√3]

10.   [34√2 –89 (½ √3) + 55(1/3)√3]

n. : ± [$f_{(n-1)}$ √2 - $f_{(n+1)}$ (½ √3) + $f_{(n)}$ (1/3)√3 ; (+) for n *even*, (-) for n *odd*

BLUE RAY, ever increasing:

	Small segment increment:			Large segment increment:

S: $\quad\quad \sqrt{(\sqrt{4}-\sqrt{3})}$ $\quad\quad\quad\quad$ S: $\quad\quad \sqrt{(\sqrt{4}-\sqrt{3})}$

T: $\quad \sqrt{2}$ $\quad\quad\quad\quad\quad\quad\quad\quad$ T: $\quad \sqrt{2}$

L: $\quad \sqrt{2} - \sqrt{(\sqrt{4}-\sqrt{3})}$ $\quad\quad\quad$ L: $\quad \sqrt{2} - \sqrt{(\sqrt{4}-\sqrt{3})}$

....

1. $\quad$ T: $\quad \sqrt{2}$ $\quad\quad\quad\quad\quad\quad\quad$ T: $\quad \sqrt{2}$

2. $\quad$ T+S: $\quad \sqrt{2} + \sqrt{(\sqrt{4}-\sqrt{3})}$ $\quad\quad$ T+L: $\quad 2\sqrt{2} - \sqrt{(\sqrt{4}-\sqrt{3})}$

3. $\quad\quad\quad\quad 2\sqrt{2} + \sqrt{(\sqrt{4}-\sqrt{3})}$ $\quad\quad\quad\quad\quad 3\sqrt{2} - \sqrt{(\sqrt{4}-\sqrt{3})}$

4. $\quad\quad\quad\quad 3\sqrt{2} + 2\sqrt{(\sqrt{4}-\sqrt{3})}$ $\quad\quad\quad\quad\quad 5\sqrt{2} - 2\sqrt{(\sqrt{4}-\sqrt{3})}$

5. $\quad\quad\quad\quad 5\sqrt{2} + 3\sqrt{(\sqrt{4}-\sqrt{3})}$ $\quad\quad\quad\quad\quad 8\sqrt{2} - 3\sqrt{(\sqrt{4}-\sqrt{3})}$

6. $\quad\quad\quad\quad 8\sqrt{2} + 5\sqrt{(\sqrt{4}-\sqrt{3})}$ $\quad\quad\quad\quad\quad 13\sqrt{2} - 5\sqrt{(\sqrt{4}-\sqrt{3})}$

7. $\quad\quad\quad\quad 13\sqrt{2} + 8\sqrt{(\sqrt{4}-\sqrt{3})}$ $\quad\quad\quad\quad\quad 21\sqrt{2} - 8\sqrt{(\sqrt{4}-\sqrt{3})}$

8. $\quad\quad\quad\quad 21\sqrt{2} + 13\sqrt{(\sqrt{4}-\sqrt{4})}$ $\quad\quad\quad\quad\quad 34\sqrt{2} - 13\sqrt{(\sqrt{4}-\sqrt{3})}$

... $\quad\quad$ .... $\quad\quad\quad\quad\quad\quad$ ....

... $\quad\quad$ .... $\quad\quad\quad\quad\quad\quad$ ....

n $\quad\quad f_{(n)}\sqrt{2} + f_{(n-1)}\sqrt{(\sqrt{4}-\sqrt{3})}$ $\quad\quad f_{(n+1)}\sqrt{2} - f_{(n-1)}\sqrt{(\sqrt{4}-\sqrt{3})}$

BLUE RAY, ever decreasing:

1. $\quad$ (T:) $\quad [\quad \sqrt{2}$

$\quad$ - (L:) $\quad - [\quad \sqrt{2} - \sqrt{(\sqrt{4}-\sqrt{3})}]$

$\quad$ = (S:) $\quad = [\quad\quad\quad \sqrt{(\sqrt{4}-\sqrt{3})}]$

2. $\quad [\quad \sqrt{2} - \sqrt{(\sqrt{4}-\sqrt{3})}]$

$\quad - [\quad\quad\quad \sqrt{(\sqrt{4}-\sqrt{3})}]$

$\quad = [\quad \sqrt{2} - 2\sqrt{(\sqrt{4}-\sqrt{3})}]$

3. $\quad [\quad\quad\quad \sqrt{(\sqrt{4}-\sqrt{3})}]$

$\quad - [\quad \sqrt{2} - 2\sqrt{(\sqrt{4}-\sqrt{3})}]$

$\quad = [- \quad \sqrt{2} + 3\sqrt{(\sqrt{4}-\sqrt{3})}]$

4. $\quad [\quad \sqrt{2} - 2\sqrt{(\sqrt{4}-\sqrt{3})}]$

$\quad - [- \sqrt{2} + 3\sqrt{(\sqrt{4}-\sqrt{3})}]$

$\quad = [\quad 2\sqrt{2} - 5\sqrt{(\sqrt{4}-\sqrt{3})}]$

5. $\quad [- \sqrt{2} + 3\sqrt{(\sqrt{4}-\sqrt{3})}]$

$\quad - [\quad 2\sqrt{2} - 5\sqrt{(\sqrt{4}-\sqrt{3})}]$

$\quad = [- 3\sqrt{2} + 8\sqrt{(\sqrt{4}-\sqrt{3})}]$

6. $\quad [\quad 2\sqrt{2} - 5\sqrt{(\sqrt{4}-\sqrt{3})}]$

$\quad - [- 3\sqrt{2} + 8\sqrt{(\sqrt{4}-\sqrt{3})}]$

$\quad = [\quad 5\sqrt{2} - 13\sqrt{(\sqrt{4}-\sqrt{3})}]$

7. $\quad [- 8\sqrt{2} + 21\sqrt{(\sqrt{4}-\sqrt{3})}]$

8. $\quad [13\sqrt{2} - 34\sqrt{(\sqrt{4}-\sqrt{3})}]$

9. $\quad [- 21\sqrt{2} + 55\sqrt{(\sqrt{4}-\sqrt{3})}]$

10. $\quad [34\sqrt{2} - 89\sqrt{(\sqrt{4}-\sqrt{3})}]$

$n^{th}: \pm [f_{(n-1)}\sqrt{2} - f_{(n+1)}\sqrt{(\sqrt{4}-\sqrt{3})}] ; \quad$ (+) for n *even* , (-) for n *odd*

PURPLE RAY, ever increasing:

Small segment increment:		Large segment increment:	
S:	$\sqrt{4}$ - $\frac{1}{2}\sqrt{6}$	S:	$\sqrt{4}$ - $\frac{1}{2}\sqrt{6}$
L:	$\frac{1}{2}\sqrt{6}$	L:	$\frac{1}{2}\sqrt{6}$
T:	$\sqrt{4}$	T:	$\sqrt{4}$

....

1.	T:	$\sqrt{4}$		T:	$\sqrt{4}$
2.	T+S:	$2\sqrt{4}$ - $(\frac{1}{2}\sqrt{6})$		T+L:	$\sqrt{4}$ + $(\frac{1}{2}\sqrt{6})$
3.		$3\sqrt{4}$ - $(\frac{1}{2}\sqrt{6})$			$2\sqrt{4}$ + $(\frac{1}{2}\sqrt{6})$
4.		$5\sqrt{4}$ - $2(\frac{1}{2}\sqrt{6})$			$3\sqrt{4}$ + $2(\frac{1}{2}\sqrt{6})$
5.		$8\sqrt{4}$ - $3(\frac{1}{2}\sqrt{6})$			$5\sqrt{4}$ + $3(\frac{1}{2}\sqrt{6})$
6.		$13\sqrt{4}$ - $5(\frac{1}{2}\sqrt{6})$			$8\sqrt{4}$ + $5(\frac{1}{2}\sqrt{6})$
7.		$21\sqrt{4}$ - $8(\frac{1}{2}\sqrt{6})$			$13\sqrt{4}$ + $8(\frac{1}{2}\sqrt{6})$
8.		$34\sqrt{4}$ - $13(\frac{1}{2}\sqrt{6})$			$21\sqrt{4}$ + $13(\frac{1}{2}\sqrt{6})$
...		......			......
...		......			......
n		$f_{(n+1)}\sqrt{4}$ - $f_{(n-1)}(\frac{1}{2}\sqrt{6})$			$f_{(n)}\sqrt{4}$ + $f_{(n-1)}(\frac{1}{2}\sqrt{6})$

PURPLE RAY, ever decreasing:

1.   T: $[\ \sqrt{4}\ \ \ \ \ \ ]$
     - L: - $[\ \ \ \ \ \ (\frac{1}{2}\sqrt{6})\ ]$
     = S: $[\ \sqrt{4}\ -\ (\frac{1}{2}\sqrt{6})]$

2. $[\ \ \ \ \ \ \ \ \ (\frac{1}{2}\sqrt{6})\ ]$
     - $[\ \sqrt{4}\ -\ \ (\frac{1}{2}\sqrt{6})\ ]$
     = $[\ -\ \sqrt{4}\ \ +\ 2(\frac{1}{2}\sqrt{6})]$

3. $[\ \sqrt{4}\ -\ \ \ \ \ (\frac{1}{2}\sqrt{6})]$
     - $[\ -\ \sqrt{4}\ +\ \ 2(\frac{1}{2}\sqrt{6})]$
     = $[\ 2\sqrt{4}\ -\ \ \ 3(\frac{1}{2}\sqrt{6})]$

4. $[\ -\ \sqrt{4}\ +\ \ 2(\frac{1}{2}\sqrt{6})\ ]$
     - $[\ 2\sqrt{4}\ -\ \ \ 3(\frac{1}{2}\sqrt{6})]$
     = $[\ -\ 3\sqrt{4}\ +\ \ \ 5(\frac{1}{2}\sqrt{6})]$

5. $[\ 2\sqrt{4}\ -\ \ \ 3(\frac{1}{2}\sqrt{6})\ ]$
     - $[\ -3\sqrt{4}\ +\ \ 5(\frac{1}{2}\sqrt{6})\ ]$
     = $[\ 5\sqrt{4}\ -\ \ \ 8(\frac{1}{2}\sqrt{6})\ ]$

6. $[\ -3\sqrt{4}\ +\ \ 5(\frac{1}{2}\sqrt{6})\ ]$
     - $[\ 5\sqrt{4}\ -\ \ \ 8(\frac{1}{2}\sqrt{6})]$
     = $[\ -8\sqrt{4}\ +\ \ 13(\frac{1}{2}\sqrt{6})\ ]$

7. $[\ 13\sqrt{4}\ -\ \ \ 21(\frac{1}{2}\sqrt{6})\ ]$

8. $[\ -\ 21\sqrt{4}\ +\ \ 34(\frac{1}{2}\sqrt{6})]$

9. $[\ 34\sqrt{4}\ -\ \ \ 55(\frac{1}{2}\sqrt{6})]$

10. $[\ -\ 55\sqrt{4}\ +\ \ 89(\frac{1}{2}\sqrt{6})\ ]$

n^{th}: $\pm\ [\ f_{(n)}\sqrt{4}\ -\ f_{(n+1)}(\frac{1}{2}\sqrt{6})$; (+) for n *odd*, (-) for n *even*

ORANGE RAY, ever increasing:

| Small segment increment: | | Large segment increment: | |

S: (2/3)√3 -   (½)√2   S: (2/3)√3 -   (½)√2
L: √2 + √(√4-√3) -   (2/3)√3   L: √2 + √(√4-√3) -   (2/3)√3
T: (½)√2 + √(√4-√3)   T: (½)√2 + √(√4-√3)

·· .

1.   { T:}   (½)√2 + √(√4-√3)   {T:}   (½)√2 + √(√4-√3)
2. {T+S} (0)(½)√2 + √(√4-√3) + (2/3)√3   {T+L:} (3/2)√2 + 2 √(√4-√3) - (2/3)√3
3.   1(½)√2 + 2√(√4-√3) + (2/3)√3   (4/2)√2 + 3√(√4-√3) - (2/3)√3
4.   1(½)√2 + 3√(√4-√3) + 2(2/3)√3   (7/2)√2 + 5√(√4-√3) - 2(2/3)√3

5.   2(½)√2 + 5√(√4-√3) + 3(2/3)√3   (11/2)√2+ 8 √(√4-√3)- 3(2/3)√3
6.   3(½)√2 + 8√(√4-√3) + 5(2/3)√3   (18/2)√2+ 13√(√4-√3)- 5(2/3)√3
7.   5(½)√2 + 13√(√4-√3)+ 8(2/3)√3   (29/2)√2+ 21√(√4-√3) – 8(2/3)√3
8.   8(½)√2 + 21√(√4-√3)+ 13(2/3)√3   (47/2)√2+ 34√(√4-√3) -13(2/3)√3

9   13(½)√2+ 34√(√4-√3)+ 21(2/3)√3   (76/2)√2+55√(√4-√3) – 21(2/3)√3

....
n   f $_{(n-2)}$ √2 + f $_{(n)}$ √(√4-√3) + f $_{(n-1)}$ (2/3)√3   l $_{(n)}$(½ √2) + f $_{(n+1)}$ √(√4-√3) – f $_{(n-1)}$ (2/3)√3

ORANGE RAY, ever decreasing:

1.   T: [(½)√2 + √(√4-√3)   ]   2.   [√2 + √(√4-√3) -   (2/3)√3]
     -L: - [√2 + √(√4-√3) - (2/3)√3]   - [- (½)√2   +   (2/3)√3]
     = S: [- (½)√2   + (2/3)√3]   = [(3/2)√2 + √(√4-√4)- 2(2/3)√3]

3.   [- (½)√2   + (2/3)√3]   4.   [(3/2)√2 + √(√4-√3) -   2(2/3)√3]
       - [(3/2)√2 + √(√4-√3) - 2(2/3)√3]   - [-(4/2)√2 - √(√4-√3) +   3(2/3)√3]
       = [- (4/2)√2 - √(√4-√3) + 3(2/3)√3]   = [(7/2)√2 + 2√(√4-√3) -   5(2/3)√3]

5.   [- (4/2)√2 -   √(√4-√3) + 3(2/3)√3]   6.   [(7/2)√2 + 2√(√4-√3) -   5(2/3)√3]
       - [(7/2)√2 + 2√(√4-√3) - 5(2/3)√3]   - [-(11/2)√2 - 3√(√4-√3) +   8(2/3)√3]
       = [-(11/2)√2 - 3√(√4-√3) + 8(2/3)√3]   = [(18/2)√2 + 5√(√4-√3) -   13(2/3)√3]

7.   [- (11/2)√2 - 3√(√4-√3) + 8(2/3)√3]   8.   [(18/2)√2 + 5√(√4-√3) -   13(2/3)√3]
       - [(18/2)√2 + 5√(√4-√3) - 13(2/3)√3]   - [-(29/2)√2 - 8√(√4-√3) +   21(2/3)√3]
       = [- (29/2)√2 - 8√(√4-√3) +21(2/3)√3]   = [(47/2)√2 +13√(√4-√3) -   34(2/3)√3]

9.   [- (29/2)√2 - 8√(√4-√3) + 21(2/3)√3]   10.   [(47/2)√2 + 13√(√4-√3)- 34(2/3)√3]
       - [(47/2)√2 +13√(√4-√3) - 34(2/3)√3]   - [-(76/2)√2 - 21√(√4-√3) +55(2/3)√3]
       = [- (76/2)√2 - 21√(√4-√3)+ 55(2/3)√3]   = [(123/2)√2+ 34√(√4-√3)- 89(2/3)√3]

ORANGE RAY, ever decreasing, ... cont'd

11.　　$[- (76/2)\sqrt{2} - 21\sqrt{(\sqrt{4}-\sqrt{3})} + 55(2/3)\sqrt{3}]$
　　　$- [(123/2)\sqrt{2} + 34\sqrt{(\sqrt{4}-\sqrt{3})} - 89(2/3)\sqrt{3}]$
　　　$= [- (199/2)\sqrt{2} - 55\sqrt{(\sqrt{4}-\sqrt{3})} + 144(2/3)\sqrt{3}]$

12.　　$[(123/2)\sqrt{2} + 34\sqrt{(\sqrt{4}-\sqrt{3})} - 89(2/3)\sqrt{3}]$
　　　$-[- (199/2)\sqrt{2} - 55\sqrt{(\sqrt{4}-\sqrt{3})} + 144(2/3)\sqrt{3}]$
　　　$=[(322/2)\sqrt{2} + 89\sqrt{(\sqrt{4}-\sqrt{3})} - 233(2/3)\sqrt{3}]$

....　　　　..................

....　　　　..................

....　　　　..................

n. :　　$\pm [l_{(n)} \sqrt{2} + f_{(n-1)} \sqrt{(\sqrt{4}-\sqrt{3})} - f_{(n)} (2/3)\sqrt{3}] ;$

(+) for n *even*, (-) for n *odd*.

Appendix III.
Table of Fractal Number Trigonometric Ratios
i.e., calculation of the solutions of the system of equations for the Circle Family Pair in Figure 10b

[1]-[¼]=[¾]: $\sqrt{¾}=½\sqrt{3}$			
[2]-[¼]=[7/4]: $\sqrt{7/4}=½\sqrt{7}$	[2]-[4/4]=[1]: $\sqrt{1}$		
[3]-[¼]=[11/4]: $\sqrt{[11/4]}=½\sqrt{11}$	[3]-[4/4]=[2]: $\sqrt{2}$	[3]-[9/4]=[¾]: $\sqrt{[¾]}=½\sqrt{3}$	
[4]-[¼]=[15/4]: $\sqrt{[15/4]}=½\sqrt{15}$	[4]-[4/4]=[3]: $\sqrt{3}$	[4]-[9/4]=[7/4]: $\sqrt{[7/4]}=½\sqrt{7}$	
[5]-[¼]=[19/4]: $\sqrt{[19/4]}=½\sqrt{19}$	[5]-[4/4]=[4]: $\sqrt{4}$	[5]-[9/4]=[11/4]: $\sqrt{[11/4]}=½\sqrt{11}$	[5]-[16/4]=[1]: $\sqrt{1}$
[6]-[¼]=[23/4]: $\sqrt{[23/4]}=½\sqrt{23}$	[6-[4/4]=[5]: $\sqrt{5}$	[6]-[9/4]=[15/4]: $\sqrt{[15/4]}=½\sqrt{15}$	[6]-[16/4]=[2]: $\sqrt{2}$
[7]-[¼]=[27/4]: $\sqrt{[27/4]}=½\sqrt{27}$	[7-[4/4]=[6]: $\sqrt{6}$	[7]-[9/4]=[19/4]: $\sqrt{[19/4]}=½\sqrt{19}$	[7]-[16/4]=[3]: $\sqrt{3}$
[8]-[¼]=[31/4]: $\sqrt{[31/4]}=½\sqrt{31}$	[8]-[4/4]=[7]: $\sqrt{7}$	[8]-[9/4]=[23/4]: $\sqrt{[23/4]}=½\sqrt{23}$	[8]-[16/4]=[4]: $\sqrt{4}$
[9]-[¼]=[35/4]: $\sqrt{[35/4]}=½\sqrt{35}$	[9]-[4/4]=[8]: $\sqrt{8}$	[9]-[9/4]=[27/4]: $\sqrt{[27/4]}=½\sqrt{27}$	[9]-[16/4]=[5]: $\sqrt{5}$
[10]-[¼]=[39/4]: $\sqrt{[39/4]}=½\sqrt{39}$	[10]-[4/4]=[9]: $\sqrt{9}$	[10]-[9/4]=[31/4]: $\sqrt{[31/4]}=½\sqrt{31}$	[10]-[16/4]=[6]: $\sqrt{6}$
[11]-[¼]=[43/4]: $\sqrt{[43/4]}=½\sqrt{43}$	[11]-[4/4]=[10]: $\sqrt{10}$	[11]-[9/4]=[35/4]: $\sqrt{[35/4]}=½\sqrt{35}$	[11]-[16/4]=[7]: $\sqrt{7}$
[12]-[¼]=[47/4]: $\sqrt{[47/4]}=½\sqrt{47}$	[12]-[4/4]=[11]: $\sqrt{11}$	[12]-[9/4]=[39/4]: $\sqrt{[39/4]}=½\sqrt{39}$	[12]-[16/4]=[8]: $\sqrt{8}$
[13]-[¼]=[51/4]: $\sqrt{[51/4]}=½\sqrt{51}$	[13]-[4/4]=[12]: $\sqrt{12}$	[13]-[9/4]=[43/4]: $\sqrt{[43/4]}=½\sqrt{43}$	[13]-[16/4]=[9]: $\sqrt{9}$
[14]-[¼]=[55/4]: $\sqrt{[55/4]}=½\sqrt{55}$	[14]-[4/4]=[13]: $\sqrt{13}$	[14]-[9/4]=[47/4]: $\sqrt{[47/4]}=½\sqrt{47}$	[14]-[16/4]=[10]: $\sqrt{10}$
[15]-[¼]=[59/4]: $\sqrt{[59/4]}=½\sqrt{59}$	[15]-[4/4]=[14]: $\sqrt{14}$	[15]-[9/4]=[51/4]: $\sqrt{[51/4]}=½\sqrt{51}$	[15]-[16/4]=[11]: $\sqrt{11}$
[16]-[¼]=[63/4]: $\sqrt{[63/4]}=½\sqrt{63}$	[16]-[4/4]=[15]: $\sqrt{15}$	[16]-[9/4]=[55/4]: $\sqrt{[55/4]}=½\sqrt{55}$	[16]-[16/4]=[12]: $\sqrt{12}$
[17]-[¼]=[67/4]: $\sqrt{[67/4]}=½\sqrt{67}$	[17]-[4/4]=[16]: $\sqrt{16}$	[17]-[9/4]=[59/4]: $\sqrt{[59/4]}=½\sqrt{59}$	[17]-[16/4]=[13]: $\sqrt{13}$
[18]-[¼]=[71/4]: $\sqrt{[71/4]}=½\sqrt{71}$	[18]-[4/4]=[17]: $\sqrt{17}$	[18]-[9/4]=[63/4]: $\sqrt{[63/4]}=½\sqrt{63}$	[18]-[16/4]=[14]: $\sqrt{14}$
[19]-[¼]=[75/4]: $\sqrt{[75/4]}=½\sqrt{75}$	[19]-[4/4]=[18]: $\sqrt{18}$	[19]-[9/4]=[67/4]: $\sqrt{[76/4]}=½\sqrt{67}$	[19]-[16/4]=[15]: $\sqrt{15}$
[20]-[¼]=[79/4]: $\sqrt{[79/4]}=½\sqrt{79}$	[20]-[4/4]=[19]: $\sqrt{19}$	[20]-[9/4]=[71/4]: $\sqrt{[71/4]}=½\sqrt{71}$	[20]-[16/4]=[16]: $\sqrt{16}$
[n]-[¼]=(4n-1)/4: $½\sqrt{(4n-1)}$	[n]-[4/4]=(n-1): $\sqrt{(n-1)}$		

The Fractal Number Trigonometry

[7]-[25/4]=[¾]: $\sqrt{[3/4]}=½\sqrt{3}$
[8]-[25/4]=[7/4]: $\sqrt{[7/4]}=½\sqrt{7}$
[9]-[25/4]=[11/4]: $\sqrt{[11/4]}=½\sqrt{11}$

Appendix IV.
Index of Bracketed Data

p. 31 $\{\sqrt{1}, \sqrt{2}, [\sqrt{2}]^{-1}\}.$ [17]

p. 31 $\{\sqrt{1}, \frac{1}{2}[\sqrt{5} + \sqrt{1}], \frac{1}{2}[\sqrt{5} - \sqrt{1}]\}.$ [18]

p. 46 $AB \equiv [1]^{1/n}, \quad n = 1, 2, 3, \ldots,$ [1]′

p. 61 $AB \equiv [p/q]^{[m/n]}$ [1′]

p. 61 $\{\sqrt{1}, (\frac{2}{3})\sqrt{3}, (\frac{1}{2})\sqrt{3}\}$ [16′]

p. 61 $[x + \frac{1}{2}AB]^2 + [y]^2 = [AB]^2$
$[x - \frac{1}{2}AB]^2 - [y]^2 = [AB]^2$
$y = 0$ [16+3]

p. 62 $(\frac{2}{3})\sqrt{3}, \sqrt{1}, \frac{1}{2}\sqrt{3}; \sqrt{2}, \sqrt{1}, \frac{1}{2}\sqrt{2}; \frac{1}{2}[\sqrt{5}+\sqrt{1}], \sqrt{1}, \frac{1}{2}[\sqrt{5}-\sqrt{1}].$ [17]

p.64 $f(x) = a_0 + a_1 x + a_2 x^2 + a_3 x^3 + \ldots + a_n x^n + a_{(n+1)} x^{(n+1)} + \ldots ,$ [18]

p. 64 $f(\mathbf{x})$, and $f(\mathbf{x} \pm \mathbf{x}_0).$ [19a]

p. 64 $\mathbf{f(x)}$, and $\mathbf{f(x \pm x}_0)$ [19b]

p. 64 $f(x + iy) = u(x, y) + iv(x, y) = G(u + iv)$ [19c]

Appendix V.
The Serendipitous Discovery of the Fractal Integer Arithmetic

By Robert L. Powell, Sr.

Director, The G. R. Lomanitz Laboratory of Visual Mathematics of The Practical Science Institute, Houston, TX and Greensboro, NC

"In mid-1978, a professional artist colleague, Professor John Biggers of the Texas Southern University School of Art, presented Powell with an urgent and insistent aesthetic inquiry in connection with the qualitative embodiment of the quantitative relationship that Leonardo da Vinci had referred to as the Divine Proportion.

Biggers desired to discover some of the special gestalt which this relation endowed in the canonical works of ancient Artists and Architects, as admired by da Vinci. The inquiry led to a several week duration study, in Biggers' Art School, of the geometry rules which seemed to govern and to guide the composers of the ancient compositions in Art and Architecture. The several weeks-long study enabled Biggers to establish a new and profound personal and professional connection with the canon of geometry rules which seemed to guide and govern the ancient composers' works of Art and Architecture.

The code so admired by da Vinci was recognized to be merely the most elegant of a hierarchy of only three codes which governed and guided the holistic integrity of all the canonical works of ancient Sacred Geometry.

A meta-mathematical study and synthesis of the three codes, by Powell, led to the recognition that the three codes formed the hierarchy of expression of an ancient Theorem of Euclid; a Theorem of Euclidean Geometry which calibrates the entire Euclidean plane in sets of Cartesian coordinates; the hierarchy of coordinates calibrates the plane, with positional notation, in terms of ($\pm$) powers of the three fractal number integer vectors: $\sqrt{3}$; $\sqrt{2}$; and $\frac{1}{2}[\sqrt{5}+\sqrt{1}]$."

Appendix VI.
About the Authors

ROBERT LEE POWELL, SR.
Inventor, Mathematician and Physicist
Born: Kerns, Texas of African-American heritage
Education: Fisk University, Nashville, TN   Degree: B.S. & M.S., Physics
Co-discoverer of the Holographic Interferometry Non-Destructive Testing method

A veteran teacher at Lowell Technical Institute, Lowell, MA; Oakland University, Rochester, MI, and Texas Southern University, Houston, TX and has worked for a variety of corporations.

This veteran physics professor is the co-discoverer of the esoteric field of work—Holographic Interferometry Non-Destructive Testing, which is a beautiful and appropriate engineering and technological use of the hologram-making process.

His knowledge and expertise also extend to some of the traditions of African art in the application of mathematical ratios and principles of Euclidean Geometry. His knowledge and lectures on the subject have influenced and changed the lives of many artists including the late Dr. John Thomas Biggers, and aspiring artists in colleges and universities in the U.S. and around the world.

In his 1997 visual mathematics installation "In This House," a part of the Project Row Houses in Houston Texas, he delivered a multi-level system of ancient mathematical teachings from a diverse mixture of sub-cultures. His cutting-edge mathematics uses the rules and tools of Euclid, and the geometrical roots of a quadrivium: Sacred art, Sacred architecture, Modernity's physics, and Modernity's biology.

Photo credit: Bright Moments – in loving memory

VANDORN HINNANT III

Vandorn Hinnant's art career spans a period of fifty years. His 1980 inclusion in the North Carolina Museum of Art group exhibition "Afro American Artists: NC, USA" was sufficient impetus for him to dedicate his life to creating imagery for the edification of the human mind/spirit.

He studied Art Design at NC A&T State University and studied sculpture at University of North Carolina-Greensboro.

Today, the artist spends most of his time working on large-scale sculpture commissions for municipalities and universities as well as sculptural forms for museum exhibits and corporate venues.

Most of Hinnant's career has been spent creating works that are organized into the following series: Organic Abstractions, The Portal Series, Geometric Abstractions, The Totemic Sculpture Series, and the MetaPhysical Geometries Series.

His works of art are in numerous corporate collections and in a great many private collections across North America. Some works are also in Africa and in Europe.

http://vandornhinnant.com
http://lightweavings.com

ROBERT L. POWELL, JR.

Robert L. Powell, Jr. served as an Associate Professor at NC A&T State University in the Civil, Architectural and Environment Engineering Department, Greensboro, North Carolina focusing on Building Design, Building Metrics, Community-Based Design and Sustainable Development. He also has been responsible for design and project management for various architectural projects. His undergraduate degree is in Architectural Engineering from Stanford University and he received a Master of Architecture (M. Arch.) degree from Massachusetts Institute of Technology.